D0742839

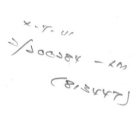

Cocoa

TROPICAL AGRICULTURE SERIES

The Tropical Agriculture Series, of which this volume
forms part, is published under the editorship of
D. Rhind, CMG, OBE, BSc, FLS, FIBiol.

Cocoa

Third Edition

G. A. R. Wood
previous editions by **D. H. Urquhart**

Longman
London and New York

Longman Group Limited   London

Associated companies, branches and representatives
throughout the world

Published in the United States of America
by Longman Inc., New York

First published 1955
Third impression with new statistics 1956
Second edition 1961
Third edition 1975
Reprinted 1978, 1979, 1980
ISBN 0 582 46667 9

Library of Congress Catalog Card Number: 73-94322

Printed and bound in Great Britain by
William Clowes (Beccles) Limited, Beccles and London

Preface to the Third Edition

In his preface to the second edition, D. H. Urquhart said that a good deal of progress had been made during the five years that had elapsed since the first edition had been published. Twelve years have passed since the second edition appeared and progress since then has been much greater and over a much wider field. During the 1950s research was concentrated largely on pests and diseases but in the 1960s there were important results in the fields of plant breeding and agronomy; this third edition presents the new information now available. I have been greatly helped by those who have contributed chapters: P. F. Entwistle of the Unit of Invertebrate Virology, Oxford, who has written the chapter on insects and cocoa; D. B. Murray, Head of the Cocoa Research Unit, Trinidad, the chapters on botany and on shade and nutrition; and A. J. Smyth of FAO, Rome, on soils. In addition to these three, I have been assisted by two of my colleagues, R. A. Lass and A. P. Williamson, who have been largely responsible for the chapters on diseases and marketing respectively. I am most grateful to all these and to many other people who have commented on various chapters.

In this edition the data is presented in metric units or in both metric and British units where appropriate; a table of conversion factors is provided on p. 280. Ceylon has become Sri Lanka and New Guinea has become Papua–New Guinea: the new names are used.

I would like to add a tribute to the work that D. H. Urquhart did during the 1950s in encouraging potential cocoa-growing areas to plant cocoa. His work is beginning to be expressed in cocoa production. It is hoped that this edition will provide guidance to growers and extension workers who are tackling the problems that arise in new and old cocoa-growing areas.

November 1973   G. A. R. Wood

Contents

Acknowledgements

The publishers are grateful to the following for the use of illustrations: British Ecological Society for Figs. 9.1, 9.2 and 9.3 from 'Light as an Ecological Factor', *Br. Ecol. Sym. No. 6*, by D. B. Murray and R. Nichols (1966); Cadbury Schweppes Ltd., for Figs. 9.4 and 9.5 from *Cocoa Growers' Bulletin*, No. 9, by D. B. Murray (1967); The Cocoa, Chocolate and Confectionary Alliance, for Fig. 12.2, from the Report on Cacao Research 1945–51 in Trinidad; Commonwealth Agricultural Bureau, for Fig. 11.5, from *Commonwealth Institute of Biological Control Technical Bulletin*, Vol. 11, by F. J. Simmonds (1969); Director of the Department of Agriculture and Fisheries, Papua and New Guinea, for Figs. 11.3, 11.6 and 11.17, from *Papua and New Guinea Agricultural Journal*, Vol. 13, by J. J. H. Szent-Ivany (1961); Dr. Enrique Ampuero, for Plates 48–51; Dr. P. Silva, for Fig. 11.13; Food and Agriculture Organization of the United Nations, for Fig. 4.1, from *FAO Soils Bulletin No. 5—Selection of Soils for Cocoa*; Institut Français du Café et du Cacao, for Plates II, III, IV and V; Inter-American Institute of Agricultural Sciences, Turrialba, for Plates 52 and 53; Longman Group Ltd., for Figs. 11.1, 11.2, 11.4, 11.7, 11.8, 11.9, 11.10, 11.11, 11.12, 11.14, 11.15, 11.17, 11.18, from *Pests of Cocoa*, by P. F. Entwistle; Macmillan and Co. Ltd., London and Basingstoke, for Fig. 2.2, from *Cacao* by Van Hall (Fig. 18); Ministry of Agriculture, Lands and Fisheries, Trinidad, for Fig. 7.1, from *A Guide to Cocoa Rehabilitation under the Cocoa Subsidy Scheme*, published by the Cocoa Board, 1954; Smithsonian Institution Press, for Fig. 2.1, from *Cacao and its Allies: A Taxonomic Revision of the Genus Theobroma*, by Jose Cuatrecasas (Fig. 1). Contributions from the United States National Herbarium, Vol. 35, published 1964.

Plates

Colour

Black and white

Chapter 1

Introduction

The cocoa tree belongs to the genus *Theobroma*, a group of small trees which occur in the Amazon basin and other tropical areas of South and Central America. There are over twenty species in the genus but only one, *Theobroma cacao*, is cultivated widely.

The origin of the cocoa tree is thought to have been in the headwaters of the Amazon basin and 'it may be assumed that in early times a natural population of *Theobroma cacao* was spread throughout the central part of Amazonia–Guiana, westward and northward to the south of Mexico; that these populations developed into two different forms geographically separated by the Panama isthmus' (Cuatrecasas, 1964). This is a recent theory on the origin of the two main types of cocoa, Criollo and Forastero, the former being found to the north and west of the Andes, the latter in the Amazon basin.

The early history of cocoa cultivation remains a matter for speculation but, at the time of the discovery of America, cocoa was being consumed by the Aztecs and there is no doubt that cocoa had already been cultivated for several centuries. It is also known that the Mayas had grown cocoa in Mexico and Guatemala. The type of cocoa grown by the Aztecs and the Mayas was Criollo and there is no indication that Forastero cocoa was cultivated until the Spaniards started to extend the planting of cocoa in South America. The reason for this may well lie in the fact that Criollo can give a palatable drink with little or no preliminary fermentation whereas Forastero cocoa requires several days' fermentation.

At the time of the conquest of Mexico in the early sixteenth century cocoa beans were not only used in the preparation of a drink, but were also used as currency, sometimes for payment of tribute to Aztec overlords, and were used in various rituals and for medicinal purposes (Thompson, 1956). As a drink the beans were first roasted and then ground and mixed with maize and annatto, chilli or some other spice. This mixture was made into a thick drink, called chocolatl.

The Spaniards found that cocoa beans could be made into a more palatable drink when mixed with sugar and in this form it became popular in Spain whence it spread to other European countries. As a result the Spaniards introduced sugar to Mexico and cocoa to some of their other possessions. In this way the cultivation of cocoa spread to Trinidad and other islands and countries around the Caribbean, thus supplying the growing markets in Europe. The drinking of chocolate spread from Spain to France and England in the seventeenth century but it was an expensive drink consumed, for instance, in the chocolate houses of the aristocracy in London. It remained a luxury in England during the seventeenth and eighteenth centuries largely due to the high duties which were imposed on beans and on 'every gallon of chocolate'. At the beginning of the nineteenth century the duty on beans was as high as 1s 10d per pound so it is not surprising that imports at that time amounted to only 200–250 tons and half of this was used in making cocoa for the Navy. Duties were reduced progressively until 1844 when they reached the level of one penny per pound for beans from British possessions, the duty on foreign cocoa beans being reduced to the same level in 1853. These reductions helped to increase the market for cocoa products and by 1850 consumption in Great Britain had risen to 1,400 tons.

During the seventeenth and eighteenth centuries the cocoa tree was introduced to many countries beyond the Caribbean. It was taken westward to the Philippines and the East Indies and thence to Ceylon (now Sri Lanka) and it was moved through Brazil and on to São Tomé and Fernando Po in the early part of the nineteenth century. It was from these islands that cocoa was introduced to the rest of West Africa towards the end of the century. Despite all these movements the production of cocoa remained centred on the Caribbean and South America until near the end of the nineteenth century. Production in 1850 has been estimated at 16,000 tons, most of which was produced in Ecuador, Venezuela, Trinidad and the State of Para in Brazil. The same four countries remained the leading producers until near the end of the century, when production had risen to 100,000 tons.

While the substantial reduction of duties enabled the price of cocoa products to be reduced, technical developments were equally important in the subsequent increase in consumption. The first of these was the development of a press to remove part of the cocoa butter from the beans which led to the manufacture of cocoa powder and chocolate as we know them today. The cocoa press was invented by Van Houten in 1828. Its use produced a more palatable cocoa powder containing only half the original fat and the cocoa butter produced was added to a mixture of roasted cocoa beans and sugar, making a product which was easy to mould. Thus it became possible to manufacture an eating chocolate and this was first marketed about the middle of the nineteenth century. Prior to this development the 'chocolate' or cocoa drink containing all the cocoa butter in the bean had been the only product; the popularity of this drink declined but its use lingered on until relatively recently in the cocoa supplied to the Royal Navy and it is still a popular drink in Colombia and Mexico.

The second technical development was the invention of milk chocolate in which milk solids are mixed with cocoa mass and sugar; it was invented by M. D. Peter of Vevey in Switzerland and was first produced in 1876. It is the growth of consumption of milk chocolate in a variety of forms that has been the most striking feature of the cocoa and chocolate industry during the present century and this product forms the backbone of the chocolate industry today.

In 1900 cocoa production throughout the world amounted to 100,000 tons, of which 15 per cent came from Africa, where the largest producer was São Tomé: in 1971/72 world production rose to over 1,500,000 tons, of which nearly 1,200,000 tons or 80 per cent was produced in Africa (Table 1.1).

Table 1.1   *Growth in cocoa production in certain countries, 1850–1970*

Country	Production (000 long tons)			
	1850	*1900*	*1950*	*1970*
Ecuador	5·5	23	32	60
Venezuela	5·4	9	17	18
Brazil	3·5	18	153	179
Trinidad	1·7	12	9	4
São Tomé		17	8	10
Ghana		1	262	386
Nigeria			110	303
Ivory Coast			56	177
Cameroon		1	47	110
World total	18	115	803	1,481

SOURCES: 1850 figures from Gordian Essays on Cocoa 1936; the other figures from Gill and Duffus (December 1972).

Table 1.2   *World consumption of cocoa beans*

Year	Consumption (000 long tons)
1900	102
1910	203
1920	376
1930	488
1940	700
1950	781
1960	927
1970	1,336

SOURCES: 1900–1920 Imports: Gordian Essays on Cocoa 1936.
1930 onwards Grindings: Gill and Duffus (1972).

There was a similarly rapid increase in consumption between 1900 and 1940 (Table 1.2). Since then production and consumption have nearly doubled but the rate of increase has been far from steady. Recent developments are dealt with in more detail in a later chapter; suffice to say here that consumption increased at 4 per cent per annum during the 1960s. This figure is close to the estimate of 3·7 to 3·9 per cent per annum made by the Food and Agriculture Organisation for the period 1961–63 to 1975. The FAO forecasts a lower rate of growth of 2·2 to 2·7 per cent for the following ten years but this means that consumption could reach 2·2 to 2·4 million tons by 1985 (FAO, 1967).

The bulk of the cocoa grown in West Africa is Amelonado, a variety of Forastero. Amelonado cocoa produces chocolate with a good flavour but without any of the additional flavours associated with the 'fine' types. These additional flavours are not usually considered necessary for the production of milk chocolate. The growth in consumption of milk chocolate has, therefore, been paralleled by the increasing production of Amelonado cocoa. While the production of Amelonado cocoa has increased enormously so that it now provides about 80 per cent of the world's supplies, the production of Criollo and Trinitario types has only increased a little so that the 'fine' types of cocoa with their various characteristic flavours now form a much smaller proportion of world production. Production of Criollo and Trinitario cocoas cannot be equated with production or sales of 'fine' types as the latter depend not only on the type of trees which are grown but also on the way in which the cocoa is prepared. While the 'fine' types as a whole command premiums over West African cocoas, a considerable proportion of the production of Venezuela and Trinidad, for instance, is sold at no premium at all. The share of the world production receiving a premium would be difficult to calculate accurately, but it is certainly less than 10 per cent.

The value of world cocoa exports is given in Table 1.3 and is compared with the export earnings of certain other tropical crops. This indicates the relative importance of cocoa in world trade but the figures can only be taken as a rough guide as production and prices vary considerably from year to year.

Table 1.3   *Value of tropical export crops—1967*

Crop	Volume of exports 000 tons	Price $ per ton	Export value $m
Coffee	3,027	711	2,153
Cocoa	1,084	539	663
Tea	620	991	574
Rubber	2,375	439	839

SOURCE: FAO *Commodity Reviews and Outlook*, 1968–69.

1. Young cocoa plant showing leaf arrangement on chupon and fans.

2. Young cocoa plant showing jorquette.

4. Collections of pods, Trinidad. Two pods in the back row of pentagona type.

3. Trunk of cocoa tree with flowers, cherelles and mature pods.

5. Criollo type pods.

6. Prolific flowering on a tree in Ghana.

7. A cocoa tree about seven years old grown from a fan cutting.

8. A chupon arising from a fan cutting.

General characteristics of the cocoa crop

In the New World cocoa is cultivated on estates and on smallholdings but estates of 20 hectares (50 acres) and upwards are the customary units. In Trinidad estates are relatively small, few exceeding 120 to 160 hectares (300 to 400 acres) but in Brazil and Ecuador some much larger estates have been established. In practically all cases the estates were originally planted by individual or family owners, but there are a few places in the West Indies and Latin America where cocoa has been planted on a plantation scale. There are, for instance, the large plantings of cocoa on land held by the United Fruit Company in Costa Rica, the cocoa being planted after the failure of bananas owing to Panama disease; these plantings were subsequently split into smaller individual units. There have also been some large-scale plantation developments in Ecuador.

In Africa cocoa is grown almost entirely on smallholdings and it is usually stated that each farm is very small. It is true that individual plantings representing one year's clearing are generally small—less than 1 hectare (3 acres)—but there is little relationship between such plantings and the size of one farmer's holding. Polly Hill (1962) has made it clear that the size of farmers' holdings and the manner in which the farms are held and become established vary enormously; any generalisation on this point would be unwise. In Nigeria the cocoa survey conducted by the Nigerian Cocoa Marketing Board produced data which showed that the area of cocoa held by most farmers was only 0·6 hectares; on the other hand 'the bulk of the cocoa produced comes off farms with a good deal more cocoa land', such farms being more than 2·5 hectares in extent (Galletti *et al.,* 1956). The same general picture is probably true of the other main cocoa-growing countries in West Africa—Ivory Coast and Cameroon—though there is less information available. In West Africa as a whole, therefore, the size of cocoa farms varies considerably but the majority of farmers hold relatively small acreages and farmers with more than 8 hectares are exceptional and their holdings cover a relatively small proportion of the land.

There are certain exceptions to this general picture. In West Cameroon several cocoa plantations were started by German plantation companies before the First World War, but only one of these plantations exists today; many of the others were cut out as a result of heavy losses from black pod disease. More recently cocoa has been planted quite extensively on some plantations in the Congo, and in Nigeria cocoa is one of the crops grown on some State Agricultural Development Corporations' estates. In addition there are a few European-owned estates in the Ivory Coast. In the Far East cocoa is a relatively new crop and is being grown on company-owned plantations in Malaysia and on privately owned estates and smallholdings in Papua–New Guinea.

Cocoa has not, so far, become a plantation crop. There are several reasons for this; first, cocoa is not grown on a large scale in those countries where plantation agriculture is widespread; second, plantations are most successful where they grow a crop which requires heavy capital expenditure or will give better yield and higher prices through skilled management. Cocoa does not require heavy capital expenditure on processing equipment; it can be processed on any scale. Furthermore, skill in processing cannot guarantee a higher price for cocoa. To be competitive with smallholdings, a cocoa plantation must achieve far higher yields and the means of doing this are beginning to be seen.

References

Cuatrecasas, J. (1964) 'Cacao and its allies. A taxonomic revision of the genus *Theobroma*', *Contrib. US Nat. Herb.*, **35** (6), pp. 379–614.
Food and Agriculture Organisation (1967) *Agricultural commodities—Projections for 1975 and 1985*, Rome, FAO, Vol. 1, pp. 232–43.
Galletti, R., Baldwin, K. D. S. and Dina, I. O. (1956) *Nigerian Cocoa Farmers*, Oxford University Press, p. 149.
Gill & Duffus (1972) *Cocoa Statistics, December 1972*, Gill & Duffus Group, London.
Hill, Polly (1962) 'Social factors in cocoa farming', in *Agriculture and Land Use in Ghana*, ed. B. J. Wills, Oxford University Press, pp. 278–85.
Thompson, J. E. S. (1956) 'Notes on the use of cacao in Middle America', in *Notes on Middle American Archaeology and Ethnology*, No. 128, Carnegie Institution of Washington, pp. 95–116.
Anon. (1936) *Gordian Essays on Cocoa*, Hamburg.

Chapter 2

The Botany of Cocoa

D. B. Murray   *University of the West Indies, St Augustine, Trinidad*

Origin

Cocoa, *Theobroma cacao* L., is one of some twenty-two species that constitute the genus *Theobroma*, a member of the family Sterculiaceae. The genus is indigenous to the New World and the species range from southern Mexico in the north to Brazil and Bolivia in the south. The centre of origin is considered to be the basin of the Upper Amazon. Only *Theobroma cacao* produces the cocoa of commerce, though other species are sometimes used as adulterants.

Natural habitat

The natural home of the cocoa tree is the lower tree storey of the New World evergreen tropical rainforest. In this environment the tree is subject to a high mean annual temperature with little variation, a high annual rainfall with short dry season, a fairly constant high relative humidity and a low sunlight intensity. It often stands in moving flood waters of rivers for several months of the year. Under these conditions cocoa survives, but its yield is very low; an occasional pod will, however, give sufficient seed to ensure the survival of the species.

Seed and germination

The seed of cocoa is non-resting, which means that storage for any length of time is not possible. Methods of storage, e.g. in damp charcoal, only slow down the rate of growth to allow storage over a period of four to six weeks. If not harvested at maturity, the seed may germinate within the pod while still on the tree. When taken from the ripe fruit, the seed is surrounded by a mucilaginous pulp which contains a germination inhibitor. Germination can be speeded up by removing this pulp with the testa which is thin and leathery. The testa is usually called the skin or, when dry, the shell.

The cotyledons of the seed are ex-albuminous and epigeal being taken above ground in the process of germination. Healthy seeds from ripe pods usually give a germination of 90 to 95 per cent.

Roots

The tap root of the seedling grows straight down into the ground. At a very early stage lateral roots arise in a collar just below the surface and in the mature tree it is found that most of the secondary roots have arisen within 15 to 20 cm of the surface.

Development of the roots is strongly influenced by the structure of the soil and, in particular, the water and air relations. On poorly drained soils with a high water table the tap root may go no deeper than 45 cm and terminate in a club, but where waterlogged conditions do not occur the tap root will go very much deeper.

Given the right conditions, cocoa cuttings can be made to root. It has been suggested that the absence of a tap root in a tree derived from a cutting could lead to a shorter and less productive life, but there is no evidence for this. A cutting usually develops two or three leader roots which penetrate the soil to a depth similar to that reached by a tap root and serve the similar function of anchoring the tree; the absorption of water and mineral nutrients is done mainly by the laterals. In an established planting the ultimate rootlets form a dense layer just below the decaying leaf litter on the soil. Because of this very shallow root layer, any form of tillage or surface cultivation is usually harmful to the tree.

Cocoa roots carry a mycorrhizal fungus, but little is known of the nature of the relationship.

Trunk and branches

The cocoa tree grows to a height of 8 to 10 metres, though it tends to be smaller when grown without shade. The habit of growth is characteristic and unusual. The seedling grows as an unbranched, single stem to a height of 1 to 2 metres. The terminal bud then ceases growth and three to five lateral branches develop at the same level to form the 'jorquette' (*jorqueta*) (Cook, 1916). Further increase in height is made by the development of a bud from below the jorquette into a vertical growth. This sucker, watershoot or *chupon* as it is termed, will again form a jorquette a few feet higher up, and in unpruned trees the process may be repeated a third or even fourth time. Development of the branches of the new storey usually leads to degeneration of those of the storey below. Vertical growth in cocoa is therefore sympodial.

Cocoa shows two types of branches, the first already described as the

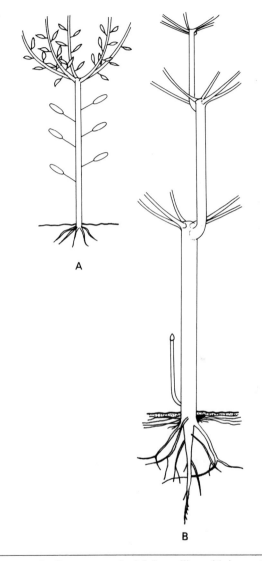

Fig. 2.1 Stem growth of cocoa tree. A. Adult seedling with jorquette and 5 fan branches. B. Older trees of 3 storeys with basal chupon.

upright and determinate chupon type, and the second the indeterminate laterals or 'fan' branches which develop at the jorquette. Besides their general direction of growth, the chupons carry their leaves in a 3/8 spiral. The fan branches have alternate leaves—hence the name. This is described as dimorphism, the juvenile, chupon growths being ortho-tropic and the mature fan branches plagiotropic.

Generally, chupons produce only chupons, except at the formation of the jorquette and, fan branches normally only give rise to fans, but may occasionally produce a chupon.

This habit of branching is important in relation to the production of plants from cuttings. Since chupons suitable for cuttings are usually available only in small numbers as compared with fans, most cuttings are taken from fan branches and their subsequent growth with low branching is quite different from that of a seedling. A fan cutting may however produce at a later stage a chupon at its base, and some planters allow this to develop at the expense of the fan part of the young tree. In this way a tree with the erect habit of growth of a seedling may be produced from a fan cutting.

Leaves

The leaves also show dimorphic characters corresponding to the differ-ent types of stem on which they arise. On the chupons, the first leaves are long-petioled and symmetrical; the petiole has a marked pulvinus or swelling at each end which allows the orientation of the leaf in relation to light. The leaves on the fan branches have shorter petioles and are slightly asymmetrical.

Production of leaves on the fan branches is by a series of 'flushes'. The terminal bud grows out rapidly producing three to six leaves. These may be pale green or various shades of red and are soft and delicate. They hang vertically but gradually 'harden' and assume, on the fan branches, their typical dorsiventral orientation. Red-pigmented leaves have their pigments masked by the development of chlorophyll while hardening and also become green. After the flush has expanded the terminal bud remains dormant for a period determined by various environmental factors and then produces another flush of growth.

Development of a new flush leads to a heavy demand on nutrients and this demand is met to a greater or less extent by translocation from the older leaves. This leads to leaf fall, hence the common description of flushing as 'change of leaf'. The extent to which the old leaves are lost when flushing occurs is a good indication of the state of nutrition of the tree. A healthy tree will produce a new flush with only a small loss of old leaves whereas heavy leaf fall indicates a shortage of mineral nutrients. In extreme cases the development of a new flush may be at the expense of all the old leaves on the tree.

The leaves carry stomata only on their under surface. Their number per unit area is markedly affected by the light intensity under which the leaf develops. Light intensity also affects the size and thickness of the leaf and its chlorophyll content, leaves under shade being larger and greener than leaves exposed to full light.

Inflorescence and flowers

Cocoa is cauliflorous, that is, the flowers and fruit are borne on the old wood of the trunk and main branches. The inflorescence arises in a leaf axil and is a very compressed dichasial cyme. This structure is sometimes revealed when stimulation by infection of the fungus, *Marasmius perniciosus* (Witches' broom disease) leads to elongation of the shoot. In the course of time, secondary thickening enlarges the site of origin to form the so-called 'cushion'. A well developed cushion may carry a large number of flowers at any one time.

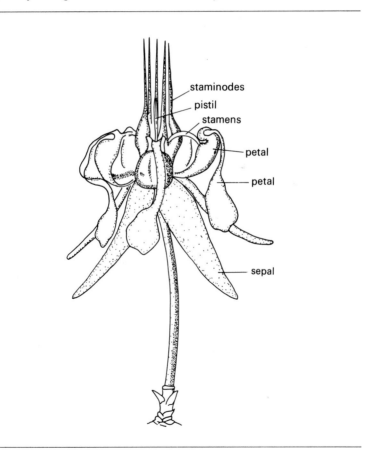

Fig. 2.2   A cocoa flower.

The long-pedicelled flower is quite regular and hermaphrodite, and has the formula K5 C5 A5 + 5 G(5), or in other words, five free sepals, five free petals, ten stamens in two whorls, only one whorl of which is fertile, and a superior ovary of five united carpels. The pink or whitish sepals are valvate in arrangement; the petals are very narrow at the base and expanded above into a cup-shaped pouch, beyond which they end in a relatively broad spatulate tip or ligule. The androecium or male part of the flower consists of five long, pointed staminodes in the outer whorl, and five fertile stamens, which, being the inner whorl, stand opposite the petals. All ten are joined at the base into a very short tube. The filaments of the stamens are bent outwards so that their anthers lie concealed in the pouched portion of the corresponding petals, whilst the staminodes stand erect and form a sort of ring fence around the style. The stamens are double by branching and therefore each has four pollen sacs. The ovary is simple, having five compartments containing numerous ovules which are arranged round a central axis in the ovary; the style is partially divided into five stigmatic lobes, which usually adhere more or less together. There is a constriction, at which the flowers absciss, at the base of the pedicel.

Pollination and incompatibility

If unpollinated the flower abscisses within twenty-four hours and a conspicuous feature of the tree at certain times is the heavy loss of flowers. A full-grown tree may produce upwards of 10,000 flowers in a year of which perhaps 10 to 50 will develop as mature fruits.

Estimates of the proportion of flowers pollinated range from 1 to 50 per cent according to the season and number of flowers opening at the time. Pollination is effected by insects of which the most important are the flying female midges of the genus *Forcipomyia* (Posnette and Entwistle, 1958; Saunders, 1958; Entwistle, 1972). A certain amount of pollination may be brought about by crawling insects such as thrips and aphids. Examinations of flowers discloses that the pollen is deposited in clusters of grains on the style and stigma, and experiments show that it may have travelled as far as 40 metres (44 yd). Most pollination occurs in the morning and artificial pollination should always be done before midday during fine weather.

The structure of the flower and the stickiness of the pollen preclude wind pollination. The absence of scent or nectar and the structure of the flower, with the anthers hidden in the pouched petals and the ring of staminodes hindering access to the stigmas, are not features to facilitate insect pollination. The midges, however, are attracted to the fleshy and purple pigmented tissue that constitutes the staminodes and 'guide lines' in the petal hoods, on which they appear to feed. In the course of feeding they present their backs, which carry numerous bristles, in turn to the stigma, style and anthers.

In the field, *Forcipomyia* is not abundant and is not easy to trap. Even though these insects are apparently present in such small numbers, pollination seldom appears to be limiting, though in Costa Rica and Ecuador, under certain conditions, fruit set may be reduced considerably, apparently through lack of pollination.

Pollination in cocoa does not necessarily imply that setting of fruit will result. Pound (1932a) showed that certain trees could not set fruit when self-pollinated or even when pollinated with pollen from certain other trees. In Trinidad, the self-incompatible and cross-incompatible tree requires pollen from a self-compatible tree to ensure fruit set, but elsewhere self-incompatible trees show cross-compatibility. The existence of self-compatible and self-incompatible trees has now been established in nearly all cocoa populations, except the major variety in West Africa, Amelonado, which is fully self-compatible (Knight and Rogers, 1955).

In most plants with incompatibility mechanisms the site of the incompatibility reaction is in the stigma or style and affects the germination and growth of the pollen tube. The mechanism in cocoa is unique in that pollen tube growth, even of incompatible pollen, is normal. Cope (1962) has shown that abnormality occurs only when the male gametes come to lie in contact with the egg and polar nuclei in the embryo sac. Fertilisation is genetically controlled by a series of alleles which show either dominance or independence interactions, and, depending on the genetic constitution of the trees involved in an incompatible pollination, either a quarter, a half or all ovules show non-fusion between the gametes. Even a quarter non-fusions result in the ovary failing to develop, and the flower falls off.

The existence of self-incompatibility is of great importance in the breeding of cocoa and in the production of hybrid seed. If one of the parents of the desired cross is self-incompatible, an isolated seed production block may be established with the self-incompatible parent interplanted with the other compatible parent. Seed collected from the self-incompatible parent is of necessity cross-pollinated. If both parents are self-incompatible, all pods may be used for seed purposes.

Fruit and cherelle wilt

After a compatible pollination the fruit starts to develop and normally reaches maturity in five to six months. The young developing fruit is known as a 'cherelle' and during the first two to three months of its development is subject to 'cherelle wilt'. This appears as a drying up and mummifying of the young pod and may account for a loss of up to 80 per cent of the developing fruit. It has been shown that this is a physiological thinning mechanism resulting from competition for water and nutrients between the young fruit, the older developing crop and also from vegetative growth. In most other crops the similar phenomenon of fruit drop results from breakdown of an abscission layer which, however, does not occur in cocoa.

After some ninety days growth, when the fruit is some 10 cm long, the production of hormones by the developing seeds ensures that cherelle wilt will not take place and the fruit will then grow to maturity unless attacked by fungal or insect pests (Nichols, 1964).

Botanically the mature fruit, usually called a pod, is best described as an indehiscent drupe. It shows a great range of variation in size, shape and colour. The length may vary from 10 cm (4 in) to 30 cm (12 in). There are basically two pod colours. Pods which when immature are green or green-white ripen to a yellow colour. Red pods darken in colour as they ripen and may show traces of orange. The number of seeds can vary from twenty to as many as fifty.

As the pod ripens, the outer layer of the integument of the seeds within produces a layer of prismatic cells which have a high content of sugar and mucilage. At full maturity these cells break down and the seeds separate easily, each surrounded by its layer of 'pulp'. The high sugar content of the pulp is of importance in the fermentation process which the seeds undergo after harvesting and is essential for the proper development of chocolate flavour.

As the ripe fruits do not open to scatter the seeds nor fall from the tree, dissemination in nature is normally by monkeys, rats and squirrels. These animals gnaw through the wall of the fruit to extract the seeds surrounded by the sweet-tasting pulp. After sucking this they drop the seed, possibly at some distance from the parent tree.

When cut across, the ripe seed which becomes the cocoa bean is found to consist mainly of the two convoluted cotyledons. These may be white in Criollo varieties or varying shades of purple in Forasteros. (The Catongo variety of Brazil is a white seeded Forastero mutant.) After fermentation and drying these beans become respectively a light cinnamon brown or a dark chocolate brown.

Parthenocarpic (seedless) pods are sometimes found in the field, particularly in Colombia and Ecuador. In all *Theobroma* species examined, the diploid number of chromosomes is 20.

Other species of *Theobroma*

The most recent treatment of the genus *Theobroma* is that of Cuatrecasas (1964) who distinguishes twenty-two species of *Theobroma* which fall into two main groups. The species in the first group have epigeal germination and subterminal vertical growth, that is, vertical growth occurs from a bud developing below the jorquette.

In the second group the species show hypogeal germination and vertical growth takes place from a bud arising above the jorquette.

Theobroma cacao belongs to the first group which also contains *Theobroma bicolor*. This species is occasionally used in South America in the production of a chocolate of inferior quality and the seeds are occasionally mixed with those of *T. cacao* as an adulterant.

Among the second group occurs *Theobroma grandiflorum* which under the name of Cupuassu is cultivated in the state of Para in Brazil. Its value lies not in the seed but in the pulp surrounding it which is used to prepare a drink and various forms of sweetmeats.

Early classifications had distinguished such species as *T. pentagonum*, the *cacao lagarto* or alligator cocoa of Guatemala; *T. leiocarpum*, a smooth-podded type from the same country; and *T. sphaerocarpum*, a nearly spherical, smooth-podded type from São Tomé. Nowadays these are all considered to be merely local variations of *T. cacao*.

Cocoa varieties and nomenclature

The earliest major exports of cocoa to Europe were made from western Venezuela where up to about 1825 only one variety was grown. Though not entirely uniform, since both red and green pods were produced, these pods were in general long and narrow, pointed, conspicuously ridged and warty. The seeds were almost round in cross section and when freshly cut were white or pale violet in colour. The cocoa produced was of good quality but the distinction should be noted between flavour, which is inherent in the bean, and quality, which is determined by the method of preparation. Thus a sample of cocoa of desirable flavour may be of poor quality through lack of fermentation or some other reason.

About 1825 introductions were made from Trinidad which gave trees with pods of quite different appearance. They were shorter and broader, not pointed, and could be quite smooth. Of more importance was the fact that their seeds when cut were purple in varying degrees. The value of these introductions was that they were hardier, more vigorous and higher yielding than the indigenous cocoa. To distinguish between the two, the older cocoa was called 'Criollo' (native) and the introductions 'Forastero' (foreign) or 'Trinitario' (of Trinidad).

The cocoa from Trinidad, though in the 'fine' trade category, was distinctly inferior to the more delicately flavoured Criollo. So it came about that on the European markets Criollo became synonymous with the best type of flavour cocoa and Forastero with a rather lower standard.

The Trinidad cocoa was botanically a very mixed group. The Spaniards first planted cocoa in Trinidad in the sixteenth century with seed believed to have been brought from Mexico. This would have been a white-seeded Criollo type. In 1727 what is described in the literature as a 'blast' occurred which virtually wiped out the cocoa in Trinidad. Whether it was a hurricane or an outbreak of disease, possibly *Ceratocystis fimbriata*, is a matter for conjecture. When the industry was being re-established cocoa was brought from eastern Venezuela, the nearest point on the mainland, and of a type quite different from the Criollo described earlier, which was being grown in western Venezuela. This cocoa was purple-seeded and when it was planted among the surviving Criollo trees, cross-fertilisation led to a hybrid population with a wide range of characters and seeds varying in colour from pure white to dark purple. It was this mixture which when introduced to Venezuela became known as Forastero or Trinitario cocoa. So in the nineteenth century the general distinction in the trade lay between Criollo cocoa with white seeds and Forastero with purple seeds. Both were what would be described today as 'fine' or 'flavour' cocoas. So too was the cocoa from Ecuador, called 'Nacional' which was again different from the cocoas of Trinidad and Venezuela.

Important developments in cocoa production began at the end of the nineteenth century. Plantings were being made in the State of Bahia in Brazil of a cocoa, supposedly introduced from Belem on the Amazon, which produced pods of a very uniform appearance. Small, nearly oval and yellow in colour they are known as 'Comun'. The seed is small and purple and the variety would fall into the Forastero class. At much the same time Ghana and Nigeria began planting introductions from Fernando Po of cocoa which had been taken from Brazil, first to São Tomé and then on to Fernando Po. This cocoa closely resembles the Bahian cocoa and is known as 'Amelonado' (melon-shaped). This Amelonado cocoa is also a purple seeded Forastero. Today by far the greatest volume of production, the so-called 'bulk' cocoa of commerce, is produced from the Amelonado Forasteros of West Africa and Brazil.

Present-day classification

Cheesman (1944) recognised the difficulty of applying the term variety to cocoa classification and preferred to speak of populations which may or may not intergrade, but at the same time show characteristics by which they can be distinguished.

He suggests the following terminology:
1. Criollo
    (*a*) Central American Criollos; (*b*) South American Criollos.
2. Forastero
    (*a*) Amazonian Forasteros; (*b*) Trinitarios.

This grouping preserves the traditional distinction between Criollo and Forastero, or 'non-Criollo' cocoas, but allows for the separation of the Forasteros into two populations, (2*a*) the Amazonian Forasteros and (2*b*) hybrid populations of recent origin.

The relationship is shown diagraphically:

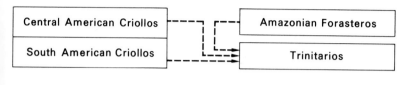

Criollo cocoas

The general pod characteristics have already been described. Cheesman distinguishes the Central and South American types as much for convenience as by conviction and considers that the original cocoas of Colombia and western Venezuela were probably indigenous to Colombia. The origin of the Central American Criollos is obscure and though sometimes assumed to be indigenous it seems more likely that they were carried up from South America by human agency. Production of Criollo cocoa is now so small that it is of little importance in world trade.

Amazonian Forasteros

Again it is perhaps useful to distinguish between the comparatively old-established ordinary cocoas of West Africa and Brazil and the newer Amazonian Forasteros originating from collecting expeditions in recent years. The Amelonado cocoa of West Africa is a very homogeneous population, resulting from a small introduction, but now represents the greater part of the world's production. Recent studies in Trinidad and elsewhere are indicating a wide range of variability in the new collections and this is proving of great value in breeding programmes.

Trinitarios

These hybrid populations resulting from natural crosses between Criollo and Forastero types show high heterogeneity. Though originally applied to the Trinidad population, the name would cover some of the cocoas of such countries as Sri Lanka, Indonesia, and Papua–New Guinea, where Criollo introductions have become crossed with later Forastero introductions.

References

Cheesman, E. E. (1944) 'Notes on the nomenclature, classification and possible relationships of cacao populations', *Trop. Agric.*, **21**, 144–59.
Cook, O. F. (1916) 'Branching and flowering habits of cacao and patashte', *Contrib. US Nat. Herb.*, **17** (8), 609–25.
Cope, F. W. (1962) 'The mechanism of pollen incompatibility in *Theobroma cacao* L.', *Heredity*, **17**, 157–82.
Cuatrecasas, J. (1964) 'Cacao and its allies. A taxonomic revision of the genus *Theobroma*', *Contrib. US Nat. Herb.*, **35** (6), 379–614.
Entwistle, P. F. (1972) *Pests of Cocoa*, Longman.
Knight, R. and Rogers, H. H. (1955) 'Incompatibility in *Theobroma cacao*', *Heredity*, **9** (1), 69–77.
Nichols, R. (1964) 'Studies of fruit development of cacao (*Theobroma cacao*), in relation to cherelle wilt. 1. Development of the pericarp, *Ann. Bot.*, n.s. **28**, 619–35.
Posnette, A. F. and Entwistle, H. M. (1958) 'The pollination of cocoa flowers', *Rep. Cocoa Conf. London 1957*, pp. 66–8.
Pound, F. J. (1932) 'Studies in fruitfulness in cacao, II. Evidence for partial sterility', *1st Ann. Rep. Cacao Res. 1931*, Trinidad, pp. 26–8.
Saunders, L. G. (1958) 'The pollination of cacao', *Cacao*, **3** (15), 1–4.

Chapter 3

Climate

The climates of some cocoa-growing countries are illustrated diagram-matically in Figs. 3.1–7 which show the rainfall, temperature and sunshine patterns of three countries in West Africa, three in the West Indies and South America and one in the Far East. These countries exhibit a wide range of climates which helps to indicate the limits for successful cocoa growing.

Seasonal pattern

The diagrams for Tafo, Ghana and Ondo, Nigeria are typical of the cocoa areas of these countries. The rainfall at these two places is 1,500 and 1,650 mm (60, 65 in) while the range for the two countries is 1,150 to 1,800 mm (45 to 70 in) in Ghana and 1,150 to 1,500 mm (45 to 60 in) in Western Nigeria. The rainfall pattern is typical of that for a latitude of 6° with two rainfall peaks separated by a brief dull, dry period in August.

The climate is similar in the main cocoa areas of Cameroon and Ivory Coast, but in West Cameroon and Fernando Po the rainfall is heavier, 2,500 mm (100 in) or more, and there is no break in the wet season.

The diagram for Uruçuca, Brazil shows an evenly distributed rainfall which is unusual at a latitude of 15°. The main difference in the weather pattern is the fall in temperature between April and October when the sun is north of the equator.

In Trinidad and Ecuador there are pronounced wet and dry seasons, the dry season lasting for four months in Trinidad and for six months in Ecuador. Under Trinidad conditions the four-month dry season makes it marginal for cocoa, and any unusually dry year leads to damage to cocoa trees. On the other hand, the longer dry season in Ecuador does not provide unsuitable conditions because it is cloudy, dull and cool, while the wet season is sunny and warm. These peculiar seasons, coupled with a high watertable, make it possible to grow cocoa and other tropical crops.

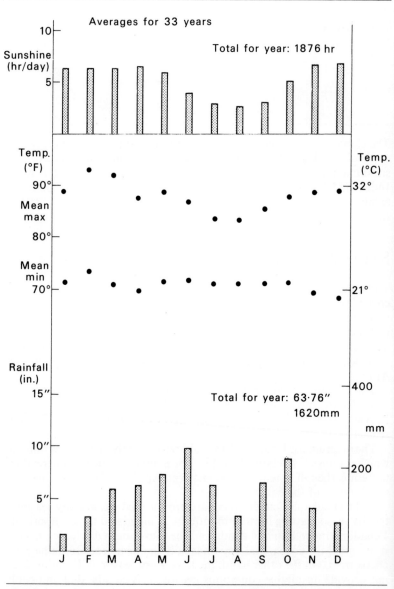

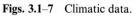
Figs. 3.1–7   Climatic data.

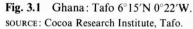
Fig. 3.1   Ghana: Tafo 6°15′N 0°22′W.
SOURCE: Cocoa Research Institute, Tafo.

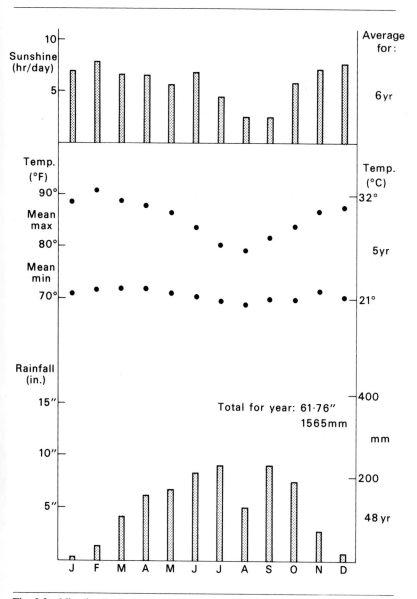

Fig. 3.2 Nigeria: Ondo 7°06′N 4°41′E, Alt.: 940 ft.

SOURCES: Rainfall and temperature—Wessel, M. 'A brief outline of the climate of the main cocoa-growing area of Western Nigeria', *Proc. Conf. Mirids and other pests of cocoa*. Ibadan 1964, pp. 90–7.
Sunshine—Agrometeorological Bulletins, Nigeria.

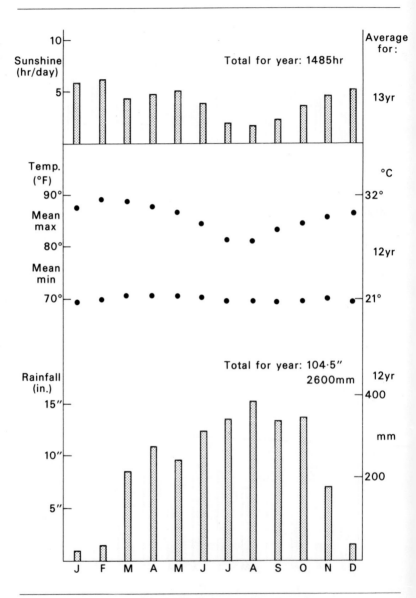

Fig. 3.3   Cameroon: Ikiliwindi 4°40′N 9°30′E, Alt.: 1000 ft.
SOURCE: Estate records.

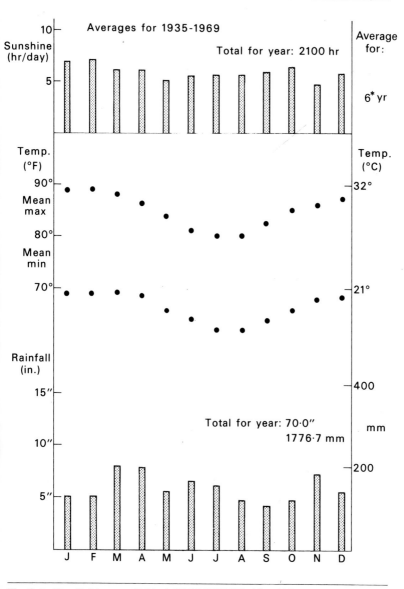

Fig. 3.4 Brazil: Uruçuca Bahia 14°36′S 39°17′W, Alt.: 129 m.

SOURCE: CEPEC. *Average for 1964–69.

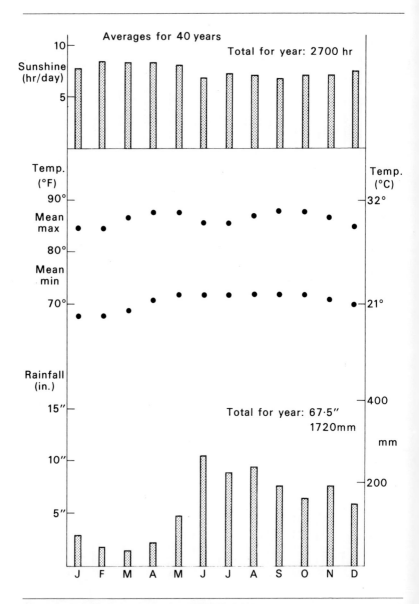

Fig. 3.5   Trinidad: St. Augustine 10°38′N 61°23′W.
SOURCE: University of West Indies, St. Augustine.

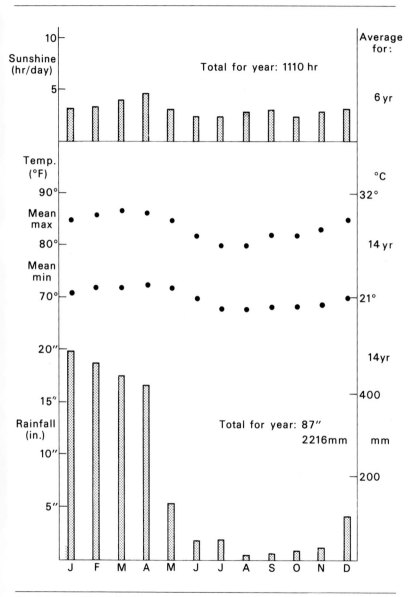

Fig. 3.6 Ecuador: Pichilingue 1°06′S 79°29′W.
SOURCE: Estacion Exptl. Trop. Pichilingue, *Informe Anual* 1960.

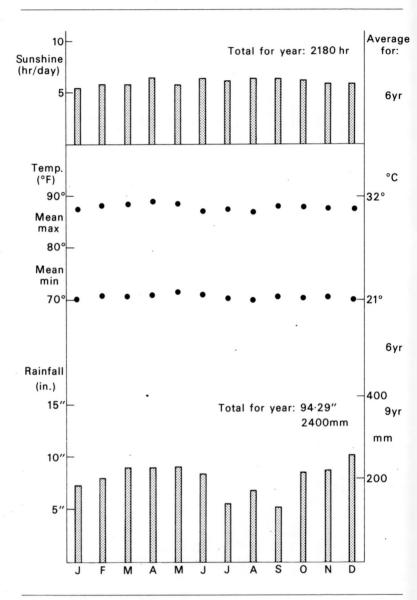

Fig. 3.7   Sabah: Quoin Hill 4°50′N 118°E.
SOURCE: Ann. Rep. Dept. Agric., Sabah 1967.

The diagram for Sabah illustrates the sort of uniform climate with little variation from month to month in rainfall, temperature and sunshine that is typical of the Far East.

From these examples, it can be seen that rainfall and its distribution is a most important climatic factor. With typical conditions of soil and temperature the dry season should not be longer than three months, and this limits the distribution of cocoa to within 20° of the equator. The basic pattern of rainfall according to latitude is for the rainfall maximum to follow the path of the zenithal sun. Thus at the equator rainfall is fairly evenly distributed throughout the year; beyond 2° from the equator there are two dry seasons of more or less equal duration but, moving away from the equator, one dry season becomes longer, the other one shorter. At 10° the long dry season may last for three to four months and beyond 15° there is only one wet and one dry season. If this pattern were regular the cocoa areas would be confined within 7° to 8° of the equator, and in fact, over 75 per cent of the world's cocoa, including all the cocoa areas in West Africa, lie within 8° of the equator.

The basic pattern of rainfall is modified by altitude, by distance from the sea and by other local conditions, so that there are many places beyond 8° where cocoa is grown. The largest cocoa area beyond this latitude is the State of Bahia in Brazil where the cocoa area lies between 13° and 18°S. Other examples from the western hemisphere are Mexico, Jamaica and the Dominican Republic whose cocoa areas lie between 18° and 20°N; in Africa there is a cocoa area at 15°S in Madagascar and in the Pacific cocoa is grown on the Fiji Islands and New Hebrides at 16° to 18°S.

Rainfall

Annual rainfall

The annual rainfall in most cocoa-growing countries lies between 1,150 and 2,500 mm (45 and 100 in). In the main cocoa area of West Africa the rainfall varies from 1,150 to 1,800 mm (45 to 70 in), but in Eastern Nigeria, West Cameroon and Fernando Po, rainfall amounts to 2,500 to 3,000 mm (100 to 120 in) per annum. In other continents rainfall in cocoa areas is generally less than 2,500 mm (100 in), some of the exceptions being parts of Mexico, the Atlantic coast of Costa Rica, the east coast of Malaya and parts of New Britain. Where rainfall is below 1,250 mm (50 in) it is probable that moisture losses from evapotranspiration will be greater than precipitation and in these circumstances cocoa will only be grown where it can be irrigated as, for instance, in the small valleys along the north coast of Venezuela, or where the ground water table is high as used to be the case in the Cauca valley of Colombia. Annual rainfall in excess of 2,500 mm (100 in) is likely to lead to greater trouble from fungus diseases, particularly black pod, owing to more humid conditions. In Papua–New Guinea, the incidence of vascular streak dieback is correlated with rainfall, occurring frequently where rainfall is above 2,500 mm (100 in).

High rainfall results in heavy leaching of the soils and in general one would expect to find poorer soils—more acid in reaction, lower in exchangeable bases—under conditions of heavy rainfall. Thus the soils on the east coast of Malaya developed from igneous rocks under a rainfall of 3,800 mm (150 in) per annum are poorer than the soils developed over similar rocks in West Africa where the rainfall is less than half that level. There are, of course, exceptions, the major one being alluvial soils whose fertility may be replenished by flooding.

It has been suggested that there is a negative correlation between rainfall and yield (Thorold, 1953, 1955). The evidence to support this is inconclusive and, while there may be indications that this is correct within a particular area, there is no sign that this correlation is true in a more general sense.

Distribution

Total annual rainfall is a less important consideration than its distribution. In the Amazon valley, rainfall is fairly evenly distributed throughout the year and it might be thought that cocoa would yield more heavily under such conditions rather than in countries with pronounced wet and dry seasons. There is no indication that this assumption is correct, and in fact, most of the world's cocoa is grown where there is a distinct dry season.

Using Thornthwaite's definition of potential evapotranspiration, it can be calculated that at Tafo there are five months in the year—four in the long dry season, one in the short—during which the losses of moisture are greater than the average precipitation. At Ondo, Nigeria, the long dry season is one month longer and the position in other areas can be seen from the diagrams. A dry season longer than four months is unusual but there are special situations, as in Ecuador, where this occurs.

Thus most cocoa-growing areas have a short mild dry season, and in Ghana cocoa is limited to those areas which receive not less than 250 mm (10 in) of rain during the five months November to March (Adams and McKelvie, 1955). In Nigeria, on the other hand, rainfall in much of the cocoa area is lower than this during the same period, at Ibadan it is only 180 mm (7 in). This contrast may be accounted for by differences in temperature, humidity or soil texture and emphasises the point that many factors have to be taken into account in assessing the suitability of an area for cocoa cultivation.

The interaction of soil texture and rainfall distribution is an important factor to be considered in assessing the suitability of an area for cocoa. Where the dry season is long or severe, soils which are deep and retentive of moisture may enable cocoa to be grown, while sandy or volcanic soils which do not retain moisture may prove suitable where rainfall is evenly distributed.

Temperature

The temperature in cocoa growing areas usually lies between a maximum of 30° to 32°C (85° to 90°F) and a minimum of 18° to 21°C (65° to 70°F). The limits of temperature are another matter. Minimum temperatures were studied by Erneholm (1948) who concluded that the lower limit for cocoa was a mean monthly minimum of 15°C (59°F) and an absolute minimum of 10°C (50°F). These lower limits will define the northerly and southerly limits of cocoa growing as well as the limit of altitude, as far as temperature is concerned.

While most cocoa areas lie below 300 m (1,000 ft), the most extensive exception is the cocoa area in East Cameroon which lies around 600 m (2,000 ft); above this level cocoa is grown in the Cauca valley in Colombia at 900 m (3,000 ft), in the Chama valley in Venezuela and in Uganda at 1,100 to 1,200 m (3,600 to 3,900 ft).

A maximum temperature for cocoa has not been determined. Hardy (1960, p. 47) has suggested a mean monthly maximum of 30°C (86°F) and this is supported by experiments in Trinidad which showed that a constant temperature of 31°C (88°F) leads to abnormal growth. Such conditions would not occur in the field where there will be diurnal variation, and it is possible that cocoa trees can withstand temperatures well above 31°C (88°F) for short periods during the day. Data from Tafo, Ghana, show a highest mean monthly maximum of 33·8°C (93°F) and at Ondo, Nigeria, it is 32·5°C (90·6°F) so that mean monthly maxima in excess of 30°C must prevail in most of the cocoa areas of West Africa.

The effect of temperature on growth and flowering of cocoa has been studied in the field and in growth rooms. Field data taken in Ghana led to the theory that flushing was suppressed when the daily maximum temperature fell below 28°C (83°F) (Greenwood and Posnette, 1950). This theory does not hold good in other places, as cocoa flushes in countries like Uganda, where temperatures do not rise to that level. Recent experiments in growth rooms have shown that flushing increases as temperature rises, but flushing will occur at low temperatures and also at constant temperature, so that temperature changes are not essential to flushing (Sale, 1967). At continuously high temperatures of 30°C (88°F) the hormone system of the tree is upset, with the result that there is a loss of apical dominance, many axillary buds developing; the leaves are smaller at high temperatures. Temperature also affects flowering, the number of flowers increasing as temperature rises.

Alvim (1966) put forward a theory that flowering is inhibited below 23°C (73°F), but further evidence (Alvim *et al.,* in press) shows that this is not correct and that competition from developing pods and soil moisture status are more important factors that can inhibit flowering.

Sunshine

The amount of sunlight falling on the cocoa tree will affect its growth and yield. This has been shown in various shade experiments which are described in the chapter on shade and manuring.

The results of these shade experiments are usually related to a certain percentage of full sunlight but it is difficult to compare results from different countries because there is little information as to the energy value of full sunlight. Thus a shade experiment in Trinidad may show that 50 per cent shade is optimum for young cocoa, but 50 per cent in Trinidad may give quite different insolation of the tree as compared with 50 per cent shade in Ghana, for instance.

Sunlight can be measured in terms of sunshine hours, and data from Campbell-Stokes recorders are fairly generally available. This, however, is not an accurate measurement of solar radiation which is measured in terms of gram-calories per square centimeter and a variety of instruments are available for this purpose. While sunshine hours and solar radiation are related, it is clear from the few comparisons that are available that the sunshine hours do not give a true measure of incident energy. The following figures illustrate this point:

Hours of sunlight: average daily figures

Month	J	F	M	A	M	J	J	A	S	O	N	D	Year
Accra, Ghana	6·8	7·2	7·0	7·2	6·8	5·2	4·5	4·5	5·7	7·1	8·1	7·6	6·45
St Augustine, Trinidad	7·6	8·2	8·0	8·1	7·8	6·7	7·2	6·9	6·5	7·0	7·0	7·1	7·3
Pichilingue, Ecuador	2·3	4·0	3·9	4·1	4·0	3·8	2·6	2·1	1·5	0·9	0·8	1·7	2·6

Mean daily solar radiation: gram cals per cm^2 (Langleys)

Month	J	F	M	A	M	J	J	A	S	O	N	D	Year
Accra, Ghana	363	436	481	485	469	387	368	358	413	451	457	399	422
St Augustine, Trinidad	385	418	478	446	470	275	388	396	394	374	308	346	390
Pichilingue, Ecuador	325	356	343	364	329	305	280	258	253	238	219	261	294

SOURCES:
For Accra—*Agriculture and Land Use in Ghana*, Wills (1962), p. 9.
For Trinidad and Ecuador—*Ann. Rep. Cacao Research*, Trinidad, 1967, p. 48.

These figures show that Accra receives more energy than Trinidad despite fewer sunlight hours and that Pichilingue, while receiving only 35 per cent of the sunlight hours when compared with Trinidad, nevertheless receives 75 per cent of incident energy.

Wind

While cocoa, in common with other tree crops, will suffer severe damage from violent winds or hurricanes, it is capable of recovering fairly rapidly if the damaged trees send up a basal chupon. Winds of this force are mercifully rare in most cocoa areas and are only likely to occur in parts of the West Indies and some Pacific Islands. Of greater importance is the sensitivity of cocoa to the steady winds which occur in the trade wind belt. The cocoa leaf has a short petiole and this can be damaged by the persistent movement caused by steady winds. Such damage can lead to defoliation. In these circumstances cocoa needs protection.

In West Africa the distribution of cocoa is influenced by the harmattan, the dry wind which blows from the Sahara between December and March. The latitude to which the harmattan penetrates varies considerably from place to place and the duration of the wind also varies from year to year. Its advent leads to a sharp drop in humidity and where the wind prevails for some weeks, as in parts of Western Nigeria, it makes the establishment of cocoa hazardous.

Other aspects of climate

Apart from affecting growth and yield, climatic conditions will also affect other aspects of cocoa growing, in particular fungal diseases and processing.

Low temperatures and high humidities, which go hand-in-hand in the wet tropics, favour the spread of fungal diseases, black pod especially. The short dry season which occurs in many parts of West Africa is of great importance in preventing the spread of black pod. This disease starts to build up during the early part of the wet season when the crop is developing, but the short dry season reduces black pod incidence. In some years this short dry period does not occur and this leads to severe losses from black pod.

The climate will also influence the methods of drying. The main crop in much of West Africa is harvested, fermented and dried during the dry season and this makes it possible to dry the beans in the sun. This has been an enormous advantage to West Africa as it enables the smallholder to prepare cocoa of good quality by simple methods of drying. The exception to this in West Africa is West Cameroon where much of the crop is harvested in the wet season and artificial drying is necessary. In Brazil and several other countries the climate is such that artificial dryers are required for at least part of the crop.

Climate, growth and cropping

To summarise, the climatic requirements for cocoa are:

1. A rainfall of 1,100 to 3,000 mm (45 to 120 in) per annum and preferably between 1,500 and 2,000 mm (60 and 80 in) with a dry season for no more than three months.
2. Temperature varying between 30° and 32°C (85° and 90°F) mean maximum, and 18° and 21°C (65° and 70°F) mean minimum.
3. No persistent strong winds.

A hot moist climate will favour growth of cocoa and in countries without a dry season cocoa will develop more quickly than in countries where growth may be stopped by low temperature or drought during certain months of the year. Countries offering conditions for continuous growth, Malaysia for instance, have an advantage over those in which growth is checked because the trees will come into bearing earlier. On the other hand there is no evidence to suggest that yields at maturity will be higher under conditions of steady growth than where there is a dry season.

The pattern of cropping of mature cocoa is clearly related to the rainfall distribution and various studies have shown that the production curve is related to the rainfall five months earlier (Alvim, 1967; Bridgland, 1953). There is also evidence that bean size is affected by rainfall during the development of the crop. There is not, however, any indication of a correlation between rainfall or any other climatic factor and actual yield, so it is not as yet possible to forecast yield or the size of a crop from a study of climatic data.

References

Adams, S. N. and McKelvie, A. D. (1955) 'Environmental requirements of cocoa in the Gold Coast', *Rep. Cocoa Conf. London 1955*, pp. 22–7.

Alvim, P. de T. (1966) 'Factors affecting flowering of the cocoa tree', *Cocoa Growers' Bull.*, 7, 15–19.

Alvim, P. de T. (1967) 'Eco-physiology of the cacao tree', *Conf. Internat. sur les Recherches Agronomiques Cacaoyères, Abidjan 1965*, pp. 23–35.

Alvim, P. de T., Machado, A. D. and Vello, F. (In press) 'Physiological responses of cacao to environmental factors', *Proc. 4th Internat. Cocoa Res. Conf., Trinidad 1972*.

Bridgland, L. A. (1953) 'Study of the relationship between cacao yield and rainfall', *Papua and New Guinea Agric. Gaz.*, **8** (2), 7–14.

Erneholm, I. (1948) *Cacao Production of South America*, Gothenburg.

Greenwood, M. and Posnette, A. F. (1950) 'The growth flushes of cacao', *J. Hort. Sci.*, **25** (3), 164–74.

Hardy, F., ed. (1960) *Cacao Manual*, Turrialba, Costa Rica, Inter-American Inst. Agric. Sciences, p. 47.

Sale, P. J. M. (1967) 'Effect of temperature on growth', *Ann. Rep. Cacao Res. 1966*, Trinidad, pp. 33–8.

Thorold, C. A. (1953) 'The control of black-pod disease of cocoa in the Western Region of Nigeria', *Rep. Cocoa Conf. London 1953*, pp. 108–15.

Thorold, C. A. (1955) 'Observations on *Theobroma cacao* in Fernando Po', *J. Ecol.*, **43** (1), 219–25.

Chapter 4

Soils

A. J. Smyth   Food and Agriculture Organisation, Rome

Basic considerations

In the world as a whole cocoa is grown successfully on many very different kinds of soil. Yet even in areas climatically well suited to the crop it is rarely safe to leave the choice of planting site to chance. Variations in climate, particularly in the volume and distribution of rainfall, dictate a change in the soil characteristics required for optimum growth and in each environmental situation the range of soils suitable for cocoa may be severely limited in both nature and extent.

In common with most plants, cocoa relies on its root system to provide anchorage and an adequate supply of mineral nutrients and water to meet the requirements of its aerial parts. Failure to meet any part of these requirements at any stage of development or during periods of adverse climate, whether occasional or seasonal, will obviously reduce the vigour of growth and may reduce yield, resistance to disease and the useful life of the tree. Furthermore, unsuitable soil conditions, particularly those of a physical nature, may not be amenable to improvement and may impose a limitation on production regardless of the quality of the planting material used or the level of husbandry employed. The selection of suitable soils is a fundamental step, therefore, in efficient cocoa farming.

The root system of cocoa shows some ability to adapt its mode of growth to soil conditions (see Fig. 4.1). Nevertheless, the failure of cocoa to compete successfully with many other tropical tree species on unfavourable soils suggests that its root system is more sensitive to factors within the soil which impede root growth, and less vigorous in obtaining nutrients and moisture when these are in short supply. This may be the primary reason why cocoa is exceptionally demanding in its soil requirements.

Root penetration and rooting volume

In describing the soil requirements of cocoa, Hardy, in particular,

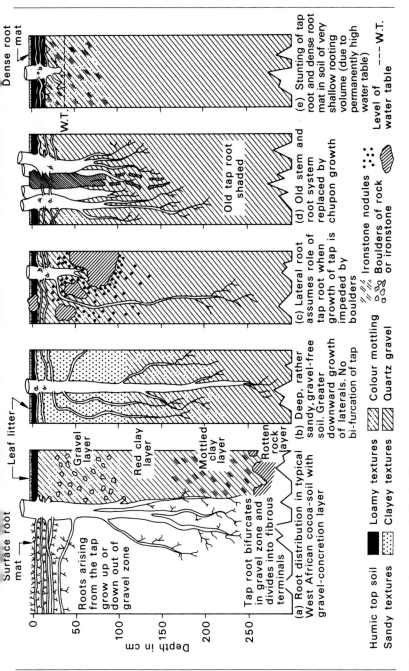

Fig. 4.1 Forms of cocoa root distribution.

has stressed the importance of 'root room' and of 'physiological depth', which he defined as the thickness of the layer of soil that is adequately aerated and structurally suitable for the unrestricted growth of roots (Hardy, 1958, 1960). Root room is important since the ability of a soil to supply moisture and nutrients depends not only on the quantity of these nutrients available in a unit volume of the soil but also on the total volume of the soil which the plant roots can exploit. Physiological depth is particularly important in the case of cocoa since other considerations, such as the need to establish a closed canopy as quickly as possible, limit the planting distance between trees and thus, to some extent, the horizontal dimensions of the rooting volume.

Therefore, in deciding the suitability of a particular site for cocoa, it would be difficult to overemphasise the importance of digging at least one pit to a depth of not less than 1·5 m (5 ft) in order to study the subsurface characteristics of the soil. Additional boreholes made rapidly with a soil auger will establish whether or not conditions observed in the pit are representative of the whole site but interpretation of the many factors which singly or in combination may inhibit root penetration requires great skill if it is to be based on the evidence of auger borings alone. Any attempt to penetrate the soil to depth will reveal the presence of rock or hardened soil close to the surface and a pit of reasonable size will show whether or not such obstacles are so extensive that the successful growth of cocoa is impossible. This simple precaution would have saved many wouldbe cocoa farmers much disappointment.

More subtle factors of the soil environment which may inhibit root growth include excess of moisture and consequent shortage of oxygen, a shortage of moisture, a shortage of nutrients or a lack of root development space between the soil particles. The last of these factors depends directly on the texture and structure of the soil, the nature of the clay minerals, and the content of large particles such as stones, gravel and ironstone concretions. The significance of all these factors is intimately related, one with another, and with environmental conditions, particularly climate. The relationships between these factors are so complex with regard to root penetration that the only simple guide lies in a study of the extent to which the roots of other plants have succeeded in penetrating the various layers of the soil, always bearing in mind that the roots of cocoa are likely to be less, rather than more, successful.

Hardy (1958) has stated that for cocoa the depth of root-penetrable soil should be at least 1·5 m (5 ft) and this may be regarded as a useful general rule. However, deeper soils are desirable where rainfall is low or badly distributed, especially if the soils tend to be rather sandy or nutritionally impoverished (Charter, 1948). Conversely, where all other aspects are particularly favourable, soils only 1 m deep may be acceptable.

9. (*above left*) Cocoa nursery, Sabah, showing polythene bags. This indicates the normal high rate of germination and uniform development.

10. (*left*) Cocoa nursery, Sabah, showing shade and general arrangement of nursery beds.

11. Cocoa seedling about 40 cm tall, ready for planting.

12. Dead seedling with bent tap root due to replanting from nursery bed to bag.

13. (*right*) Nursery trees for cuttings under Gliricidia shade. Trinidad.

14. Stem and single leaf cuttings trimmed and prepared for the rooting bin.

15. (*above right*) Effect of moisture conditions on rooting.
    (a) Overwatering and inadequate aeration: callus rods developing.
    (b) Underwatering and excessive aeration: a basal callus pad has developed.

16. Effect on leaves of abnormal conditions during rooting. **(a)** Excessive loss of moisture due to failure to maintain a high humidity causes patchiness which can be seen above; these patches are yellow. **(b)** Inadequate light intensity leads to carbohydrate starvation which is shown by yellowing areas.

17. (*right*) Potting rooted cuttings. The current practice is to raise cuttings in polythene bags with a core of rooting medium, but this photograph shows the root development after four weeks in rooting medium.

Moisture and air requirements

An adequate supply of both soil moisture and soil air is essential for healthy root development. Excess of moisture, partially filling the larger soil pores and closing the interconnecting passages, reduces the total air space in the soil and prevents a free exchange of gases between the soil and the atmosphere. This leads to a shortage of oxygen and to an accumulation of carbon dioxide in the immediate vicinity of the roots, thus inhibiting root respiration which is essential for the development of the energy required for the absorption of water and nutrients. Alvim (1959) has described how cocoa growing on rather wet soils in Colombia suffered from 'physiological drought' induced by poor soil aeration and in the classic experiments on soil aeration under cocoa conducted by Vine and others (1942, 1943), in Trinidad, levels of carbon dioxide accumulation considered to be toxic were detected within 10 cm (4 in) of the soil surface during the wet season.

On the other hand, experimental work by Alvim (1959) and Lemée (1955) has shown that, in relation to many other tropical tree crops, cocoa is exceptionally sensitive to shortages of soil moisture. The growth processes of cocoa being adversely affected by moisture short-age long before the soil is so dry that plant roots can no longer exert sufficient suction to extract any soil water (i.e. long before the soil moisture status reaches 'wilting point').

To be good for cocoa, therefore, a soil must be capable of retaining an adequate supply of available moisture during all seasons of the year, while at the same time it must have those qualities of good drainage, permitting free escape of excess water to depth, which are required to provide good soil aeration. These apparently opposed requirements are not irreconcilable, for properties of moisture retention within the soil constitute only one of several factors, within and external to the soil, which determine soil drainage qualities.

Soil texture, structure and nature of the clay

In soils composed very largely of sand, large voids very rarely form, since individual sand grains are unaffected by moisture and lack the cohesion required to form larger aggregates, or to stabilise tunnels and fissures made by animal or plant activity. The size of the pore spaces in such soils is directly related, therefore, to the average size and arrange-ment of the sand grains. If the sand grain size is fine, or very fine, which is often the case in soils formed on river levees, the spaces between grains are smaller than the growing point of cocoa roots and so cocoa root development is almost impossible. Coarse sandy soils, on the other hand, have very poor properties of moisture retention, for only a small proportion of their pore space is of capillary size. Coarse sandy soils have excellent properties of drainage and aeration but dry out too quickly to be useful for cocoa in most climates and usually have a low content of nutrients since the soluble plant foods are leached away in the drainage waters. Coarse sands, and even fine gravel of 'rice grain' size, are suitable, however, for potting mixtures for cocoa seedlings (Hardy, 1960) since moisture and nutrients can be supplied as required.

In contrast, there is great variation in the size of the pores in a soil composed mainly of clay. On wetting, very large amounts of water can be adsorbed on the relatively enormous surface area of the minute clay particles in such a soil. Depending on the nature of the clay, greater or less amounts of water can also be absorbed within the structure of the clay particles. Absorption and release of water causes a mass of clay to swell and contract as its moisture content changes. These movements lead to the formation of large structural units in the soil, between which both roots and air can freely penetrate. In certain very clayey soils (known as vertisols or grumusols and by a variety of local names) the churning and cracking action is sufficient to disrupt plant roots and such soils are unsuitable for cocoa. Although the total amount of water in a clay soil may be very large, much of it is retained by forces greater than can be exerted by plant roots and when clay soils are wet their aeration is usually very poor.

Best suited to cocoa are soils in which the 'fine-earth' is composed of 30 to 40 per cent of clay, about 50 per cent of sand and 10 to 20 per cent of silt-sized particles. In areas potentially suited to cocoa the finer particles of these soils are very often aggregated to form large, very stable, particles of about coarse sand size. Such soils possess the desirable characteristics of free drainage and good aeration associated with coarse sandy soils, while retaining properties of relatively high moisture retention associated with their content of very small particles.

In many soils, soil texture changes with increasing depth. While the texture of the surface layers is important it often varies too widely within short distances to serve as a useful criterion for distinguishing soils of differing suitability for cocoa. More consistent guidance can be obtained by comparing textures in soil layers lying between depths of about 25 and 50 cm (10 to 20 in). Very marked changes of texture, from sand to clay or vice versa, over a short vertical distance are very undesirable since the movement of moisture between layers of strongly contrasting texture tends to be restricted. In potential cocoa growing areas, however, such marked changes in texture are likely to be encountered only in soils forming in recent alluvium.

Large particles in the soil, such as stones, gravel and concretionary ironstone are undesirable in that they occupy soil volume while contributing little or nothing to the supply of nutrients and moisture. In West African cocoa areas, in particular, very large quantities of gravel are often present within the soil. Their significance in individual soils is not easy to judge. They cause bifurcation and poor development of individual roots, especially the tap root (Charter, 1948; see Fig. 4.1) and, in excessive quantities, can prevent root penetration altogether. It may be said that any appreciable quantity of gravel in the top 25 to 30 centimetres of the soil is undesirable and that, at greater depths, gravel in quantities exceeding about 40 per cent of the volume of any single soil layer is likely to make a soil unsuitable for cocoa.

Topsoil structure

Good structure in the surface layer of the soil is particularly important in relation to the growth of cocoa. A large majority of the feeding roots of cocoa tend to proliferate within the top 10 to 20 cm (4 to 8 in) of most soils, so that conditions favourable to root development are especially necessary in this layer. It is also apparent that conditions in the surface layers which favour or impede the free exchange of soil air with the atmosphere will exert a controlling influence on aeration in the soil layers below. The relatively large soil aggregates which form the desirable 'crumb structure' in the uppermost layers of highly productive soils are mainly cemented by organic compounds derived from the breakdown of organic matter and secreted by microfauna. The individual soil crumbs are rich in organic matter and in plant nutrients and the whole 'crumb layer' forms a storehouse of plant food. The organic cements, however, are much less stable than the compounds which cement aggregates at greater depth and the soil crumbs can easily be destroyed by exposing the soil surface to heavy rainfall and direct sunshine. Fine materials derived from the breakdown of crumbs tend to clog the connecting passages in the surface of the soil and, in some soils, a surface crust may form. This action has an adverse effect on moisture penetration and on soil aeration, and is detrimental to the vigorous growth of cocoa.

Undisturbed soils under natural forest in the humid tropics usually possess a relatively thick surface layer of crumb-structured material, rich in organic matter and plant nutrients. This layer undoubtedly contributes to the success of cocoa planted through natural forest or on land very recently cleared from forest. Clearing of the forest and cultivation leads to the rapid destruction of the valuable crumb layer. Protection of the crumb layer is an important aspect of soil management in cocoa husbandry and provides an important reason for establishing a closed canopy in the cocoa farm as early as possible. Even under well established cocoa, however, removal of the thick layer of leaf litter almost invariably reveals a rather compact soil surface with little development of crumb material. This may explain some of the difficulty experienced in attempting to replant cocoa on old cocoa farms and suggests that more attention should be given to the use of mulches under cocoa.

Soil colour

Differences in soil colour are due, in part, to differences in the degree of hydration and oxidation of iron compounds. Therefore, although the colour of a soil does not affect its drainage or aeration, it does provide a very useful guide to the moisture conditions which prevail in the soil and thus to the selection of soils for cocoa.

The least hydrated, most oxidised, iron compounds are reddest in colour and soils that have a bright red matrix colour can be assumed to have excellent aeration and are rarely, if ever, saturated with water. Orange-brown, brown and yellow colours arise through hydration of the iron compounds and reflect slightly inferior aeration and longer periods of high moisture status in the soils. Dull grey, blue and even green colours in the soil develop under anaerobic conditions when the layer of soil concerned is subject to prolonged periods of flooding.

A mottled colour pattern, with bright 'rusty-orange' patches of re-oxidised iron set in a dull coloured matrix, is produced in soil layers which are subjected to marked seasonal changes in moisture status. In moderately clayey and clayey soils, the presence of such mottled layers provides an accurate indication of the maximum seasonal rise of the groundwater table. In coarser textured soils, complete re-oxidation may be very rapid and absence of a mottled layer is then no guarantee that the water table does not approach the surface at some period of the year. Many tropical soils include clayey layers that are brightly mottled in red, brown, yellow and white. The bright coloration of these layers indicates present aerobic conditions and they should not be confused with the dull-coloured matrix of presently poorly drained soil layers.

Factors external to the soil

Most tropical soils are freely draining and downward movement of water is restricted only if the lower soil horizons are already saturated. This is most likely to occur in areas of high rainfall and in low topographical sites where the groundwater may approach, or rise above, the surface at some period of the year.

Cocoa tolerates short periods of complete flooding but if the water-table remains close to the surface for prolonged periods anaerobic conditions prevent the formation of deep roots. Unless properties of moisture retention in the upper soil layers are very favourable, cocoa planted under such conditions is liable to die of drought in dry years when the watertable falls below the reach of its roots. Fluctuations in the level of the watertable are usually less rapid and less pronounced in sites close to major rivers which flow all the year round. Here it may be possible to select areas for cocoa planting with a fair degree of certainty that the trees, even though they may be dependent on a very shallow, dense root mat, will only be left high and dry in very exceptional years.

Physical soil requirements under different climatic conditions

A soil which, by virtue of its content of well-aggregated fine particles, possesses good moisture retention, good drainage and good aeration, can be regarded as physically ideal for cocoa under any climatic condition in which the crop will grow. However, the extent to which soil physical characteristics can be allowed to depart from this ideal or, in other words, the range of soil physical characteristics which can be regarded as acceptable for cocoa, varies with differences in climate and, in particular, with differences in the volume and distribution of rainfall.

In areas where rainfall is very heavy throughout the year, as in parts of Papua–New Guinea and the Cameroons, good root aeration is a primary consideration. Here, exceptionally well ventilated soils on gently sloping sites high above the reach of groundwater are most suitable for cocoa. Clayey textures are acceptable only when good aeration is preserved by stable aggregation and the clay minerals have low properties of expansion. Under continuous rainfall the physical properties of coarse sandy soils are desirable but nutrient loss from these soils due to leaching is likely to be rapid. Where there is little or no risk of drought and where the use of mineral fertilisers is economically possible, sandy soils may be accepted as marginally suited to cocoa, especially if they are well supplied with organic matter. Organic matter content is especially important in the soils of high rainfall areas, for in sandy soils organic matter serves as a storehouse for moisture and plant nutrients while in finer textured soils it serves to improve topsoil structure and thus aeration (Amson and Lems, 1963).

In areas of moderate to low rainfall, and especially where the dry season is fairly pronounced, as in much of the cocoa growing area of West Africa, problems of moisture supply are more severe than those of aeration and sandy soils with excessively free drainage are to be avoided. In Western Nigeria, for example, it was concluded that a minimum of 25 per cent clay plus silt was required in the fine earth fractions of soil layers between depths of 25 to 50 cm (10 to 20 in) to support healthy cocoa (Smyth and Montgomery, 1962).

So long as rainfall is adequate, upper slope sites are to be preferred for cocoa planting because root growth is less likely to be impeded by groundwater and because lower humidity in the air discourages most insect and fungal pathogens. However, if total precipitation is too low or the dry season too prolonged for healthy growth, cocoa must either be irrigated or planted in valley bottom sites where trees can draw benefit from seasonal streams and groundwater (Aubert and Moulinier, 1954). Under such conditions, soils with a relatively high content of clay in their upper horizons are essential for cocoa planting.

The choice of planting site is especially difficult in climates which combine a very heavy wet season rainfall with a severe and prolonged dry season, as in the eastern provinces of Sierra Leone. Here the benefit of dry season groundwater may be essential for healthy growth of cocoa, but excellent drainage of the surface horizons is also necessary to minimise the effects of poor aeration during periods of heavy rainfall. Permeable soils, in which rapid and deep root growth are possible, occupying terrace sites immediately above areas of prolonged flooding, are best suited to cocoa in these circumstances (Smyth, 1966).

Soil nutrient requirements

Chemical analysis of soil material provides a broad measure of the capacity of a soil to supply the nutrient requirements of cocoa on a continuing basis. Our present knowledge of the interpretation of soil analytical data is adequate to distinguish soils which can be expected to support cocoa successfully from nutritionally impoverished soils on which success is doubtful.

When cocoa planting is contemplated in areas where the crop has never been grown before, chemical analysis of the soils to determine their general suitability is essential. Elsewhere, a reasonably reliable selection of new planting sites can be made by comparing the physical characteristics of the soils with those of sites already supporting healthy cocoa. However, analytical investigation is always a wise precaution if the proposed planting site has any of the following characteristics:

1. Soils which are sandy in texture. Apart from the relatively rapid loss of nutrients through leaching in sandy soils, sandy textures frequently reflect advanced weathering and/or derivation from coarse grained rocks originally poor in plant nutrients. Furthermore, the breakdown of organic matter proceeds more rapidly in sandy soils with consequent loss of nutrients and of capacity to store both nutrients and moisture.
2. Soils which are pale in colour. Pale soil colours usually reflect excessive leaching and/or derivation from rock materials poor in plant nutrients.
3. Soils known to be derived from acid crystalline rocks, from sediments derived from such rocks, or from old alluvium. Alluvial soils in the tropics are often formed in very strongly weathered material and their nutrient status should be regarded as suspect.
4. High annual rainfall (over 2,000 mm (80 in)).
5. Evidence, in the vegetation, of recent cultivation.

Interpretation of soil chemical analysis data

The chemical determinations which are most useful in distinguishing soils which differ widely in their chemical characteristics, and thus in their ability to support cocoa, are those which provide information on the capacity of the soil to store and release nutrients (i.e. 'base or cation exchange capacity' and 'organic matter content') and those which define the extent to which this storage capacity is already filled with nutrients (i.e. 'base saturation' and pH). Determination of the 'availability' of individual plant nutrients in the soil provides useful supporting information but on their own such data are difficult to interpret reliably, especially in relation to the requirements of a tree crop, such as cocoa.

A comparison of chemical data relating only to the top 10 to 15 cm (4 to 6 in) of the soils of different sites, obtained by analysing mixed random samples from each site, is more informative of the relative suitability of these sites for cocoa than would be true of many other tree crops, because of the tendency for the feeding roots of cocoa to concentrate in this layer of the soil. If the general suitability of the soils is unknown, however, analysis of deeper soil layers is necessary, for it is essential to discover whether the chemical characteristics of these layers will favour or limit root development.

Soils which have satisfactory physical characteristics and which, on analysis, show the following chemical characteristics can be expected to support cocoa successfully, if need be without addition of mineral fertilisers:

1. Base exchange capacity in the surface layers not appreciably less than 12 me/100 grams of fine earth and in subsoil horizons not less than 5 me/100 grams.
2. Average organic matter content in the top 0 to 15 cm (0 to 6 in) of the profile not less than 3·0 per cent (1·75 per cent organic carbon).
3. Base saturation in subsurface layers not appreciably less than 35 per cent (unless base exchange capacity is exceptionally high).
4. pH in the range 6·0 to 7·5 in the surface layers and no excessively acid (pH below 4·0) or alkaline (pH above 8·0) layers within one metre of the surface.
5. The following levels of individual exchangeable bases in the top 0 to 15 cm (0 to 6 in) of the soil:
   Calcium not lower than 8·0 me/100 grams of fine earth.
   Magnesium not lower than 2·0 me/100 grams of fine earth.
   Potassium not lower than 0·24 me/100 grams of fine earth.

Failure to meet these limits does not imply that cocoa will not grow on the soil in question but does suggest that problems of nutrition, which may not be easy to cure, are likely to be encountered if planting is attempted.

The most reliable method of assessing the chemical suitability of a soil for cocoa is to make a general comparison of available data with data from corresponding depths in soils known to support cocoa successfully elsewhere. A selection of data of this kind from different parts of the world is given in Table 4.1.

The selection of soils for cocoa

With respect to soil, the suitability of potential planting sites for cocoa can be assessed separately on the basis of evidence provided by a few carefully sited deep soil pits. However, a more reliable and more precise assessment is possible if this evidence can be compared with the findings of a systematic soil survey covering the surrounding district. A soil survey provides an understanding of the genetic relationships existing between the various kinds of soil in the area, from which a reliable estimate of the relative potential of the different soils can be derived. Furthermore, overall knowledge of soil distribution can be used to ensure that the sites chosen for planting are the most suitable available.

The exact identification of suitable planting areas for cocoa usually demands very detailed soil survey (at scales of 1:10,000 or larger) which is time-consuming and expensive. Indeed, if the location of suitable planting sites for commercial cocoa is the only immediate aim, the cost of soil surveying large areas in detail cannot be justified. The required knowledge of individual soils, and of the relationships between them, can be obtained from detailed soil surveys of small carefully selected areas. If cocoa is already growing in these areas its performance on each kind of soil can be assessed. Much cheaper soil surveys at smaller scales (such as 1:50,000) will show the broad distribution of groups of soils and will indicate in which parts of the entire area the more suitable soils are most extensive. Individual sites within these more promising areas can then be studied as required, using the knowledge gained from the detailed surveys.

Soil surveys have already been carried out in many areas which are climatically suited to cocoa. Unfortunately, a wide variety of criteria and of nomenclature has been used to distinguish the kinds of soil mapped in these surveys and, unless the requirements of cocoa were specifically considered in selecting distinguishing criteria, the maps produced may be of little immediate assistance to the wouldbe cocoa farmer. Only in recent years has a concerted effort been made to correlate and standardise the names and descriptions given to different kinds of soil in differing parts of the world and the significance of soil nomenclature, especially in the tropics, remains too complex to be discussed here. Nevertheless, a published soil survey includes descriptions of the soils which have been identified, together with soil analytical

Country	Soil series	Depth of sample (inches)	Carbon %	Total nitrogen %	C/N ratio	pH	BEC me/100 g. soil	Base saturation %	Exchange bases me/100 grams fine earth					Ratios		'Available' P ppm	Total P ppm	% Clay (in fine earth <0·002 mm)
									Ca	Mg	Mn	Na	K	Ca/Mg	Ca+Mg/K			
Ghana (SOURCE: Brammer, 1962)	Wacri (developed on hornblend-biotite granodiorite)	0–1·5	5·68	0·50	11·4	7·5	45·0		34·12	7·42	1·15	0·75	0·52	4·60	79·9		550	
		1·5–8	0·84	0·10	8·6	7·1	10·3		5·99	1·25	0·18	0·12	0·12	4·79	60·3		341	10·8
		8–14	0·20	0·03	7·1	6·5	5·8		1·89	0·64	0·10	0·13	0·10	2·95	25·3		274	7·1
		14–20	0·21	0·06	3·4	6·3	9·1		2·95	1·07	0·13	0·18	0·15	2·76	26·8		318	17·3
		39–55	0·29	0·06	5·0	5·6	19·5		5·12	5·57	0·08	0·22	0·11	0·92	96·5		257	43·3
Nigeria (Western Region) (SOURCE: Smyth and Montgomery, 1962)	Iwo (developed on coarse granitic gneiss)	0–2	2·32	0·14	16·2	7·6		97·6	10·90	3·00	0·09		0·64	3·63	21·7	42		6·0
		2–6	0·92	0·08	11·8	6·1		72·0	5·20	2·30	0·06		0·35	2·26	21·4	10·5		8·0
		6–9	0·64	0·06	10·2	6·3		76·4	5·60	1·30	0·07		0·22	4·31	31·4	7·6		11·6
		13–18	0·48	0·05	8·9	6·4		77·1	7·40	1·70	0·02		0·35	4·4	26·0	5·0		34·6
		18–34	0·47	0·06	7·7	6·6		75·2	8·00	2·70	0·02		0·45	2·96	23·8	4·7		43·2
		45–62	0·25	0·04	6·4	6·7		74·9	6·60	3·20	0·05		0·45	2·06	21·8	4·5		33·6
Nigeria (Eastern Region) (SOURCE: Wessel, 1965)	(Derived from basaltic rocks)	0–6	2·50	0·27	9·3		14·2	75	7·50	1·80		0·48	0·89	4·17	10·4	5·4		42·0
		6–12	1·30	0·14	9·3		13·4	78	6·40	2·80		0·48	0·77	2·29	11·9	2·4		52·0
		12–24	0·70	0·06	11·7		8·7	77	5·60	0·50		0·37	0·26	11·20	23·5	1·6		54·0
		24–36					8·6	77	5·50	0·50		0·42	0·18	11·00	33·3	3·0		55·0
		36–48					8·0	86	6·60	0·70		0·42	0·30	9·43	24·3	2·8		55·0
Ivory Coast (SOURCE: R. Maignien, 1963)	(Derived from amphibolite)	0–4	2·63	0·36	7·3	7·7			33·0	6·38		0·11	0·68	5·17	57·9			24·7
		12–16				7·3			15·9	6·64		0·13	0·06	2·39	375·7			30·2
		24–36				7·6			16·5	21·18		0·38	0·04	0·78	942·0			
Ivory Coast (SOURCE: Vertière, 1967)	(Derived from granitic rocks)	0–8	15·4	1·60	9·6	6·9			8·28	1·66		0·22	0·09	4·99	110·3		530	33·3
		20							2·90	2·21		0·20	0·05	1·31	102·2		360	36·2
		39							0·66	0·72		0·16	0·06	0·91	23·0		740	27·1
West Indies, Trinidad (SOURCE: Maliphant, personal communication)	River Estate, sandy loam (derived from alluvium). Data is averaged from 32 or 40 samples at each depth.	0–3	1·72	0·19	9·1	6·0	11·49	100	9·71	2·07		0·22	0·14	4·69	84·1	18·8		
		3–6	1·21	0·15	8·0	5·5	10·17	90	7·56	1·84		0·20	0·11	4·11	85·5	11·5		
		6–12	1·99	0·13	7·7	5·5	9·58	77	5·97	1·59		0·16	0·10	3·75	75·6	8·5		
		12–18	0·69	0·09	7·3	5·5	7·60	72	4·33	1·35		0·18	0·10	3·21	56·8	5·9		
Sabah (SOURCE: Gillespie, personal communication)	Jarangan (developed on basalt)	0–4	5·80	0·46	12·6	7·32			27·61	2·11			0·40	13·09	74·3	17		
		4–15	5·75	0·45	12·8	7·36			20·92	2·30			0·43	9·10	54·0	17		
		15–72	1·20	0·16	7·5	6·40			11·28	0·78			0·08	14·46	150·8	1		
Western Samoa (SOURCE: Wright, 1963)	A'ana bouldery clay (derived from basalt)	0–5	9·1	1·02	9·0	6·6	47·9	100	36·8	10·2		0·20	0·70	3·6	67·1	20		
		5–15	2·9	0·32	9·0	6·4	18·0	53	6·8	3·0		0·10	0·10	2·27	98·0	80		

data, and a careful study of this information should reveal which of the soil units mapped are likely to include soils suitable for planting.

In the absence of adequately interpreted soil survey data, geological maps are helpful in identifying areas where soils suitable for cocoa are most likely to be found. The soils best suited are almost invariably associated with the rocks which are most basic in mineral composition. Basic volcanic ash, basalt lavas and plutonic rocks rich in amphibole minerals, provide especially promising soil parent materials.

Soil quality classification for cocoa

Little advantage is to be gained from developing a complex classification to compare the relative suitability of soils for cocoa. In most areas a simple grouping of the different kinds of soils into four quality classes is sufficient.

Class I   Good soils: soils which have very few, if any, characteristics likely to limit the growth and yield of cocoa; recommended for planting in preference to all other soils.

Class II   Fairly good soils: soils which, although not as suitable as the soils of Class I, suffer from no serious limitations; recommended for planting in areas where no Class I soils are available.

Class III Poor soils: soils having one or more undesirable characteristic likely to restrict, if not prevent, the growth of cocoa; not recommended for planting in their present state.

Class IV   Unsuitable soils: soils suffering from severe limitations likely to prevent the satisfactory growth of cocoa; unsuitable for planting.

In interpreting a soil survey, it might be desirable to recognise subclasses to distinguish different kinds of limitations affecting the soils of Classes II and III, some of which might be amenable to improvement by good management in the future, but such details need not concern us here.

The simple quality classification indicates immediately which soils should be considered first for planting (Class I) and also those soils, usually very shallow or extremely sandy, on which planting is not worth attempting even if they occur in the midst of an area of good soil (Class IV). The most important distinction, however, is between the soils on which planting is recommended (Classes I and II) and those on which planting is unwise (Classes III and IV).

Soils having chemical characteristics inferior to the criteria listed previously would be excluded from Classes I and II. So also would soils having a combination of textural and structural characteristics which suggest that an unfavourable balance of soil air and soil moisture would adversely affect growth in a significant part of the year. In this connection the relationship between climate and soil quality classification should be noted. A soil with sandy texture to a depth of 40 or 50 cm might be 'fairly good' (Class II) in an area having continuously high rainfall, but would be 'poor' (Class III) or 'unsuitable' (Class IV) in areas of low rainfall.

Soils of Class I can be expected to have chemical characteristics superior to those of Class II but our access to laboratory data and our knowledge of its interpretation is usually inadequate to distinguish these classes on chemical criteria alone. In distinguishing these classes account is taken, therefore, of minor differences in the clay content of the uppermost layers of the soil or, where texture is essentially similar, of colour differences suggesting differential amounts of leaching and loss of nutrients.

References

Alvim, P. de T. (1959) 'Water requirements of cocoa', *First FAO Technical Cocoa Meeting*, Accra, Ghana.
Amson, F. W. and Lems, G. (1963) *De invoed van de bodem op de kolfproduktie in een cacaoklonnen experiment* [The influence of the soils on pod production in a clonal cocoa experiment], Mededeling No. 33, Landbouwproefstation, Surinam.
Aubert, G. and Moulinier, H. (1954) 'Observations sur quelques caractères des sols de cacaoyères en Cote d'Ivoire', *L'Agronomie Tropicale*, 9, 428–38.
Brammer, H. (1962) 'Soils', in J. B. Wills, ed., *Agriculture and Land Use in Ghana*, Oxford University Press, pp. 88–126.
Charter, C. F. (1948) *Cocoa Soils: good and bad*, circular issued by West African Cocoa Research Institute, Tafo, Ghana.
Hardy, F. (1958) 'Cacao soils', *Proc. of the Soil and Crop Science Soc. of Florida*, **18**, 75–87.
Hardy, F., ed. (1960) *Cacao Manual*, Turrialba, Costa Rica, Inter-American Inst. Agric. Sciences.
Lemée, G. (1955) 'Influence de l'alimentation en eau et de l'ombrage due l'economie hydrique et la photosynthese du cacaoyer', *L'Agronomie Tropicale*, **10**, 592–603.
Maignien, R. (1963) 'Les sols bruns entrophes tropicaux', in *Sols Africains*, Vol. 8, No. 3.
Smyth, A. J. and Montgomery, R. F. (1962) *The Soils and Land-use of Central Western Nigeria*, Ibadan, Western Nigeria, The Government Printer.
Smyth, A. J. (1966) *The Selection of Soils for Cocoa*, Soils Bulletin No. 5, Rome, FAO.
Verlière, G. (1967) 'Un essai d'engrais sur cacaoyers en Cote d'Ivoire—Amelioration des rendements par la fumure minerale et rentabilité', *Conf. Internat. sur les Recherches Agronomiques Cacaoyères, Abidjan 1965*, pp. 74–81.
Vine, H., Thompson, H. A. and Hardy, F. (1942–3) 'Studies on aeration of cacao soils in Trinidad, Pts I to IV', *Trop. Agric.*, **19**, 175–80; **19**, 215–23; **20**, 13–24; **20**, 51–6.
Wessel, M. (1965) 'The cocoa-growing area of Ikom in Eastern Nigeria', Cocoa Res. Inst. Nigeria (unpublished document).
Wright, A. C. S. (1963) *Soils and Land-use in Western Samoa*, New Zealand, Bull. Soils Bur. No. 22.

Chapter 5

Planting Material

The broad botanical classification of cocoa and the evolution of the three main types—Criollo, Forastero and Trinitario—was described in Chapter 2. A new grower may ask which of the three types to plant; this will normally be determined by the type being grown in the country concerned, but the question has been considered by the agricultural departments of several potential cocoa-growing countries in recent years. With few exceptions, the conclusion has been to grow Forastero cocoa in the belief that this will give the greatest income per acre.

Forastero cocoas are the most widely grown, and the bulk of the market is provided by the Amelonados of West Africa and Brazil. As Ghana is the largest producer the price paid for its cocoa forms a base for the market, some cocoa being sold at a discount below the price for Ghana cocoa, others being sold at a premium. The premium cocoas are generally produced from Criollo or Trinitario trees; on the other hand there are countries where these types are grown but whose cocoa fetches little or no premium. In Papua–New Guinea, for instance, Trinitario cocoa is grown, but the price paid, while varying between plantations, is generally in line with that for maincrop Ghana cocoa.

The case for planting Criollo or Trinitario cocoa is not simple and requires fuller examination. Premiums paid for 'fine' cocoa depend on the production of cocoa with distinctive flavours. This involves not only the variety planted, but also the techniques of fermentation and drying. These techniques are not simply a question of following methods used in other countries producing premium cocoas; there are many variables affecting the process of fermentation, and it is impossible to define a particular set of conditions which will ensure the production of 'fine' cocoa. While some countries have managed to improve the quality of their cocoa and enter the premium market in recent years, the market remains largely tied to countries or markets which have traditionally produced 'fine' cocoa. It must also be remembered that the market for premium cocoa is a small one and could be severely affected by the addition of a few thousand tons of 'fine' cocoa.

Another disadvantage of planting premium cocoas lies in their agricultural characteristics. Criollo cocoas are less vigorous than other types; they develop more slowly and yield less. There is also a tendency for Criollo trees to be more susceptible to diseases such as canker and *Ceratostomella* wilt. Several countries, Sri Lanka and Mexico, for example, at one time grew Criollo cocoa exclusively, but Forastero types were introduced, proved more vigorous, and have virtually replaced the original Criollo. There have been selection programmes for Criollo cocoa in Mexico, Madagascar and Venezuela, but few trials have been planted by which the value of these selections can be assessed. Some of the high yielding selections made in Trinidad look like Criollo, but do not give a Criollo flavour and remain susceptible to *Ceratostomella* wilt.

There are therefore substantial difficulties to be overcome in achieving good yields and premium prices from Criollo cocoa, and this policy can hardly be recommended to potential cocoa growing countries.

Trinitario cocoas have fewer handicaps; they are more vigorous and yield better, but seedlings are extremely variable in yield and other characteristics. In Trinidad the Imperial College selections had to be propagated vegetatively as rooted cuttings to ensure their high yield potential; this is a relatively expensive process, and rooted cuttings are less easy to handle in the field than seedlings. When hybrid varieties were developed, producing seedling progeny which were more vigorous and higher yielding than the clones, there was a change from rooted cuttings to seedlings in Trinidad and other countries in the Caribbean. While these breeding programmes have produced new selections with better agricultural characters, the special flavours associated with Trinitario cocoa have tended to be lost. A new grower planting these selections cannot therefore be sure of obtaining a premium.

The advances in breeding and selection have gone further with Forastero cocoas and there is greater assurance of high yields with the more recently developed hybrids than with the other types. The preparation of good quality cocoa from Forastero types presents no problem and the price obtained should be close to that of Ghana cocoa. The case for planting Forastero types where there is a choice is strong.

The question of quality, flavour and premium is also discussed in the chapter on quality.

The need for selection

The need for reliable planting material of good performance was demonstrated in Trinidad by the Department of Agriculture many years ago. There the trees are very variable in yield and other characteristics and the low yields of many cocoa fields were shown to be due to the large proportion of trees that yielded little or nothing. As far back as 1915 trees in a field at River Estate were analysed as follows (Freeman, 1929):

Yield per annum (pods)	Percentage of trees
0–12	23·0
13–25	20·0
26–50	30·4
51–75	15·9
76–100	6·0
100 +	4·7

These figures were reinforced by the detailed survey of cocoa estates conducted by Shepherd (1937). One set of data will illustrate the findings on this subject. On the so-called chocolate soil—the epithet refers to its colour rather than its reputation for cocoa growing—a field yielding 742 kg per hectare (662 lb/acre) contained 16 per cent of unproductive tree sites and 37 per cent trees yielding less than 0·4 kg beans making a total of 53 per cent of unprofitable trees. On other soils the proportion of unprofitable trees was higher. This survey showed the need for reliable planting material. Although reference has only been made to Trinidad, the same situation was generally true of Sri Lanka and Java, where the trees were variable in type, yield and quality.

Where selected seed was used it was obtained from high-yielding trees selected on appearance or on yield records over a short period; yields were usually measured by the number of pods produced which did not take into account the great variation from tree to tree in the weight of wet cocoa per pod. Such methods of selection often gave disappointing results, either because the selected trees did not breed true, or because the seed resulted from cross-pollination, or, thirdly, because the selected tree happened to be in a particularly favourable position. For these reasons the early attempts at selection failed to provide reliable planting material, and a more scientific approach was necessary.

Selection in Trinidad

One of the first aims of the Cocoa Research Scheme at the Imperial College of Tropical Agriculture was to select trees of high yield and good bean size and to develop methods of vegetative propagation.

Pound's survey (1932) showed that not only were the trees in Trinidad very variable in the number of pods they produced, but also in the number of pods required to produce one pound of dry beans. This factor, known as 'pod value' or 'pod index', was found to vary from 6 to 22 pods per pound of dry cocoa. There was also great variation in bean size, but it was found that large pods with low pod value contained larger rather than more beans.

Therefore a standard of pod value was set in order to reduce the labour of harvesting and to select for large bean size. On the basis of this survey, selection standards were set at a pod value not greater than 7·5 and a yield of 50 pods per annum at a spacing of 3·6 by 3·6 m (12 by 12 ft). This is equal to a yield of 1 ton of dry cocoa per acre (2·5 tons/ha). Using these standards during the survey of many thousands of trees, 100 trees were selected, which became the Imperial College Selections referred to as ICS 1–100. Although these selected trees yielded heavily they could not be expected to breed true because of their genetic heterogeneity, and vegetative propagation by rooted cuttings was used to multiply the selected trees. It has subsequently been shown that cuttings from good clones yield much more heavily than open-pollinated seedlings from the same clones (Jolly, 1956).

These clones were planted in trials at River Estate and at San Juan Estate, and, considerably later, in clonal trials in various parts of Trinidad and Tobago (Jolly and de Verteuil, 1960). These trials have shown two things: first, the performance of a clone varies widely on different soils: ICS 1, for instance, yielded well on the sandy loam soils at River Estate, but ranks thirteenth in the clonal trials on thirteen sites in Trinidad and Tobago; second, only a few clones have yielded more than 0·5 ton per acre, or half the computed yield of the original tree, and many clones have given much less and are of little or no value as planting material. The best clones, yielding over 800 kg per hectare (700 lb/acre) in these trials were, in descending order of yield, ICS 40, 39, 60, 48, 43 and 95.

In 1928 witches' broom disease caused by *Marasmius perniciosus* appeared in Trinidad and as it spread it became clear that the trees in Trinidad showed susceptibility to the disease. Pound made an expedition to Ecuador in 1936 and two expeditions to the headwaters of the Amazon in 1936–37 and 1942 to look for immune or resistant varieties (Pound, 1938, 1943). He sent back a large amount of planting material, which has been of enormous value far beyond the original purpose of the expeditions and now forms the bases of selection and breeding programmes in many countries.

The material collected in the first two expeditions was sent as seed to Barbados for quarantine and subsequently moved as budwood to Trinidad where the collection of no less than 2,500 clones was established at Marper Farm in an area where witches' broom was prevalent. This collection was identified according to its origin, the selections from Iquitos, Parinari, Nanay and Morona in the Upper Amazon being given the prefices IMC, Pa, Na and Mo respectively. Some selections from Ecuador are called Scavina or SCA, but there is doubt as to the precise source of these selections, as there is no place or establishment called Scavina. The material from the third expedition was sent to Trinidad as budlings and became the 'P' clone collection.

Two of the introductions from Ecuador, SCA 6 and 12, proved to be immune or highly resistant to witches' broom disease; they also yielded well but produced beans smaller than the normal commercial standard. These two clones therefore have been crossed with some ICS clones in order to combine the resistance and yielding ability of the one with the good bean size and flavour of the other. The progeny of these crosses have yielded heavily as shown by the following data from trials at River Estate (Bartley, 1969):

Clone or cross	Mean yield for years 8–12 after planting	
	lb/acre	kg/ha
ICS 1	1,163	1,308
ICS 6	1,381	1,550
ICS 1 × ICS 6	835	938
ICS 1 × SCA 6	2,253	2,540
ICS 6 × SCA 6	2,768	3,108

This work by the Cocoa Research Staff at the Regional Research Centre was supplemented and taken further by the Cocoa Board in Trinidad. The Board's responsibility was to propagate and distribute cocoa plants, but it also maintained a breeding and selection programme from the mid 1950s. This programme was taken over by the Ministry of Agriculture when the Cocoa Board was wound up. This programme involves crossing between selections, generally of different populations, in order to obtain hybrid vigour and to combine various characteristics. The good bean size and low pod value of the ICS selections has been combined with resistance to witches' broom of SCA 6 and resistance to *Ceratostomella* wilt of IMC 67. This has been achieved in three generations by crossing and vigorous selection amongst the progeny (Freeman, in press).

Planters in Trinidad have therefore been able to plant selected material since 1945: initially the selected material was distributed as rooted cuttings, but a change to hybrid seedlings was introduced in 1957. Similar programmes and changes have taken place in other countries in the Caribbean and Central America.

Breeding work in West Africa

The first introductions of cocoa to Ghana were made by Basel missionaries (Wanner, 1962), but the introduction from Fernando Po in 1879 by the celebrated Tetteh Quashie was of much greater importance. An introduction from the same source, but at a later date, started cocoa-growing in Nigeria. The bulk of the cocoa planted in Ghana and Nigeria is, therefore, of the same Amelonado type.

As cocoa-growing spread, the Gold Coast Agriculture Department made introductions of Trinitario and Criollo types to its station at Aburi; some seedpods were released to farmers and on many farms individual trees or groups of trees can be found with characters derived from these introductions, but they remain a very small proportion—less than 5 per cent—of the total tree population.

West African Amelonado is genetically homozygous, varying little in its agricultural characteristics. When selection programmes were started in Nigeria in 1931 and in Gold Coast (Ghana) in 1937, high yielding trees of Amelonado and the other types that had been imported by the Departments of Agriculture were selected and a number of crosses were made. Before this programme had yielded any significantly improved material, the threat from swollen shoot virus disease had shown the need to enlarge the collection of cocoa planting material; in 1944 Posnette made a large introduction of new planting material from Trinidad. It comprised over 100 pods from various sources, but the most significant were the pods from Upper Amazon selections which had reached Trinidad after Pound's expedition in search of cocoa resistant to witches' broom disease. Each pod introduced was given a number prefixed by the letter T; the first fifty-nine pods resulted from open pollinations but pods numbered T60 and upwards arose from hand pollinations. The seedlings were planted out with Amelonado at Tafo and it was soon apparent that 'as a group, the Upper Amazon types were earlier coming into bearing, higher yielding and much more vigorous than other types' (Posnette, 1951). Quality assessments were made, particularly for bean size and flavour, and as a result ten Amazon types—T 60, 63, 65, 72, 73, 76, 79, 82, 85 and 87—were approved for general distribution to farmers in 1954. A further type, T12, was approved later. With the exception of T12, which was an open pollinated pod from SCA 12, the approved types were all crosses between Parinari, Nanay and Iquitos selections.

The parentage of the eleven approved types was (Toxopeus, 1964):

$$
\begin{aligned}
\text{T60} &= \text{Pa } 7 \times \text{ Na } 32 \\
\text{T63} &= \text{Pa } 35 \times \text{ Na } 32 \\
\text{T65} &= \text{Pa } 7 \times \text{IMC } 47 \\
\text{T72} &= \text{Na } 32 \times \text{IMC } 60 \\
\text{T73} &= \text{Na } 33 \times \text{IMC } 60 \\
\text{T76} &= \text{Pa } 35 \times \text{Na } 31 \\
\text{T79} &= \text{Na } 32 \times \text{Pa } 7 \\
\text{T82} &= \text{Na } 32 \times \text{Pa } 35 \\
\text{T85} &= \text{IMC } 60 \times \text{Na } 34 \\
\text{T87} &= \text{IMC } 60 \times \text{Na } 34 \\
\text{T12} &= \text{SCA } 12 \times \ ?
\end{aligned}
$$

These T types were the F_1 generation which had been planted at Tafo. Open pollinated seeds from these trees were planted in observation plots which, after approval had been given, became multiplication plots producing F_3 Amazon seed for distribution to farmers. This was the first step in the provision of improved planting material.

While the release of the approved types was in progress, a more detailed programme was begun. All the breeding material at Tafo was surveyed and the progeny of crosses between the highest yielding selections were planted in a series of trials. The highest yielding progeny proved to be crosses between Upper Amazon selections and Ghanaian selections, some of which were Amelonado, others local Trinitarios. Some of these became the Series II varieties and biclonal seed gardens were planted in the early 1960s in order to produce seed of these varieties. The Series II varieties were selected for yield and quality and are described in more detail later. Various other series have been selected for other factors: resistance to black pod, resistance to or tolerance of swollen shoot virus disease, and drought tolerance. None of the progenies from these trials has been released for planting.

It is becoming increasingly realised in Ghana that the results obtained from formal progeny trials on research stations may not provide a reliable indication as to how the same progenies will perform on farms. Emphasis is therefore now being placed on breeding and selection for factors which may lead to a reduction of crop losses.

The programme in Nigeria has followed a similar pattern; F_3 Amazon being distributed for several years and now being replaced by new varieties, some of which have been specially selected for their ability to establish well under the more severe dry season conditions that prevail in Nigeria.

The pattern of development in Ghana and Nigeria is typical of what has been happening in other cocoa countries with selections of Amazon parentage being crossed with other selections of Forastero or Trinitario origin to give high-yielding progeny.

Breeding of better varieties is obviously a continuous process and it has been suggested (Toxopeus, 1972) that in a breeding programme there should, initially, be a collection from as many different populations as possible. Within each population trees should be selected for good bean and pod values as well as disease resistance. Selections from different populations should be crossed and the crosses tested in progeny trials. Finally a number of crosses should be selected for distribution.

The success of breeding programmes depends on the techniques used and on a large collection of original material. Much of the recent advance in planting material resulted from Pound's expeditions and his collections may not yet have been fully exploited. Material from more recent expeditions has not been fully evaluated, but there remain considerable unexplored areas of the Amazon basin and surrounding areas which may well contain cocoa trees of value to future breeding programmes.

Summary of the characteristics of the main varieties

1. Amelonado

This was the original type planted in West Africa and is still predominant. While it shows some variation in yield, Amelonado cocoa is very uniform in other respects and breeds true to type. It is not a vigorous grower and in West Africa takes up to ten years to reach maximum yield; on the other hand, when mature and in optimal conditions, Amelonado will yield heavily. In the Amelonado shade and fertiliser trial at Tafo the unshaded plots receiving fertiliser yielded over 3,375 kg per ha (3,000 lb per acre) for several years. In Sabah where the growing season is virtually continuous, Amelonado cocoa develops rapidly and has given 560 kg per ha (500 lb per acre) in the third year. Under less suitable conditions, for instance, the high rainfall conditions occurring in Malaya, Amelonado compares unfavourably with other Forastero types, developing slowly, yielding poorly, and suffering severely from dieback.

Amelonado grows tall and carries most of its crop on the trunk, but this is largely a reflection of the close spacing at which it is planted. Its cropping pattern in West Africa shows a pronounced peak, usually in November.

Amelonado is less susceptible to black pod and to canker than some Amazon types, but is generally more susceptible to swollen shoot virus disease.

2. Amazon

What is described here as Amazon cocoa is also called Upper Amazon or Amazonian; the word Amazon is used for simplicity. The origin of Amazon types and their development in West Africa have already been described.

The first new planting material distributed to farmers was F_3 Amazon which has been described as

a fast-growing cocoa type relatively easy to establish, precocious and high yielding with an improved ability to withstand adverse conditions and, because of its relatively wide parental background, a good general purpose variety. It should be borne in mind, however, that with its vigour and higher yields, deficiencies or general exhaustion of soils are likely to occur. This, together with the likelihood of more severe black pod losses, will probably become the major problem connected with its use (Toxopeus, 1964).

There is now more evidence that Amazons are susceptible to high losses from black pod and cankers than Amelonado. Attacks of canker were first reported from Tafo in 1963 (Glendinning, 1964) and also from Nigeria in 1967. The significance of this escaped attention for some time, as canker due to *Phytophthora palmivora* is easily overlooked in the field and has not been associated with Amelonado cocoa in West Africa. A strong warning of danger from this disease in relation to Amazon cocoa was sounded in 1971 (Vernon, 1971).

Another problem is presented by the pentatomid bug, *Bathycoelia thalassina* in West Africa (Lodos, 1967). This bug causes less damage to Amelonado cocoa but has caused heavy losses to Amazon cocoa, probably because of the more continuous cropping pattern. Rapid growth and a shallow root system are thought to have contributed to boron deficiency in Amazon cocoa at Tafo, but this has not been shown to be a widespread problem.

3. Amazon hybrids

This term describes the next generation of planting material resulting from selection among Amazons and crossing with selections of other types.

The Series II, or 'Tafo hybrid' varieties, are crosses between in-

Table 5.1 *Ghana Series II trials: total net yields in kilogrammes dry cocoa per hectare*

Progeny	8th PTA 1955–72	9th PTA 1956–63 and 1964–72	Apedwa 1960–72	Oyoko 1960–72	Pankese 1960–72
*A WAE 5	23,010[3]				17,636
*B WBE 2	17,370[4]				8,703
*C WBE 3	18,372[4]		9,779	12,780	13,040
*D WBE 4	18,221		10,765	11,928	14,228
E WEB 1		14,519		14,503	15,301
*H WAE 11[1]		14,822	13,493	16,996	17,424
*J WEA 3[1]		14,652	10,491	18,762	
*K WBE 6[2]		11,125	8,659	13,053	
*L WEB 3[2]		13,116	8,275		
*M WEA 2		15,490	9,754		
P WA 1 Amel	14,850				
Q WA 2 Amel	15,221				
F WE 6		13,889	10,520	12,928	16,784
G WE 4	13,355			10,469	13,860

* Approved Series II hybrid. Production of B, K and L have been replaced by an Amelonado cross and the seed garden for C now produces A.
[1] Reciprocals.
[2] Reciprocals.
[3] Single plot of 16 trees.
[4] Two plots each of 16 trees.

   The figures are based on core plots from seven years after planting onwards, 4 plots of 16 trees in 8th PTA and at Apedwa and 5 plots of 16 trees in 9th PTA. Oyoko and Pankese yields are from whole plots and are based on over 100 trees each.
Data from Lockwood (unpublished).

dividual Amazons and Amelonado or local Trinitario selections. A number of crosses were made at Tafo and planted out in two progeny trials, on the results of which twelve crosses were selected for more widespread trials at Cocoa Stations. Finally, seven crosses were selected for multiplication in seed gardens. These crosses and the yields obtained are given in Table 5.1. The progeny code indicating parentage is as follows:

W—WACRI hybrid
A —Amelonado selection
B —Local Trinitario or hybrid selection
E —Amazon selection

The female parent comes first in the code letters.

The first plantings of Amazon hybrids were designed to test the breeding value of the new Amazon material. The early growth of the wide hybrids was so spectacular that it was thought that some might be exploited directly as new varieties. Seed gardens for large-scale production of the new hybrids were planned on the basis of yields from young trees and it was not until later that it was realised that some of the hybrids suffered heavier losses from black pod than did Amelonado. In general it was found that 'as a group the Upper Amazon × Amelonado hybrids are preferable to the local Trinitario hybrids in all the characters studied' (Anon, 1970). Only one outcrossed inter-Amazon hybrid (WE 6) was included in the trials and it was not considered for general release. In some respects it is the most satisfactory of these varieties, with the lowest black pod disease losses. This result is of particular significance as the highest levels of resistance to swollen shoot virus disease are found among the Upper Amazons.

In the Ivory Coast, a similar series of hybrids has given outstanding results in trials and are now being produced in seed gardens for general distribution.

In Malaysia a range of hybrids between Amazon and Amelonado or Trinitario trees have performed well in trials and seed is available from seed gardens (Chee Yau Fah, 1970). These new varieties have been shown to be vigorous and capable of yielding heavily, but they have not been planted widely enough for a detailed assessment of them to be made.

The vigour and improved yield of most new varieties of cocoa arises from hybrid vigour which is expressed when selections from different populations are crossed. The first evidence of this came from an experiment in Nigeria (Russell, 1952). A trial was laid out comparing the selfed progeny of two Trinidad selections with the progeny of crosses between the two types. In the first five years of bearing the crosses yielded three to four times as much as the selfed progeny, and over nineteen years the crosses gave 50 per cent more (Atanda and

Toxopeus, 1971). Similarly, the trials at Tafo of the Trinidad introductions showed higher yields from crosses between different Amazon
populations than crosses within them; Parinari × Nanay progeny
yielding better than Pa × Pa or Na × Na crosses.

4. Trinitario

The Trinitario type is thought to have arisen from crossing between
Forastero and Criollo types. It is heterozygous and therefore does not
breed true. The greatest variety of trees is found in Trinidad, and the
variation applies to many other factors in addition to the obvious
profusion of pod types and differences of tree yield.

This type offers great scope for selection and the programme in
Trinidad has been described. Initially, selected trees were propagated
vegetatively, but this has been largely superseded by production of
seedlings of known crosses.

In New Guinea, now the largest producer of Trinitario cocoa, there
seems to be rather less variation, and it has been possible to obtain
good yields using open-pollinated progeny of selected trees or 'clonal
seed' from the crossing of two clones (Bridgland, 1960). While these
selections have performed well in Papua–New Guinea, they have not
yet yielded well in Malaya (Haddon, 1961).

The Trinitario type as a whole produces fairly large beans but the
shell content is usually rather high at 14 to 16 per cent. The flavour is
sometimes such as to establish a premium.

5. Criollo

Criollo cocoa is usually described as delicate, lacking vigour, low-
yielding and susceptible to disease, and this was probably true of old
Criollo populations. A true Criollo, such as the Java Criollo, has red
pods which are pointed and rough, and white beans. Selection programmes have been conducted in Indonesia, Madagascar, Mexico and
Venezuela but, with the exception of Indonesia where selection work
was initiated many years ago, few positive results have been reported.
There are several Trinitario selections, ICS 39, 40, 45 and 60, for
instance, which are described as near-Criollo. ICS 39 and 40 have
yielded well in clonal trials in Trinidad, but one cannot say whether,
planted on their own, they would produce an acceptable Criollo type
bean, a potential premium cocoa. ICS 45 and 60 have yielded well in
some trials, but in Ecuador ICS 45 was badly affected by *Ceratostomella*
wilt, while ICS 60 is highly susceptible to witches' broom disease.

The most successful Criollo type selections from a Trinitario population have been the Indonesian types originating from Djati Roenggo
estate. Originally selected by Van Hall, DR 1, 2 and 38 have been
planted in Java and in East Africa and produce a premium cocoa.

Seed gardens

The object of planting seed gardens is to provide seed of known parentage and proven performance. This object is achieved by including at least one self-incompatible parent—a tree which will not set fruit with its own pollen; this parent can then be used as the female parent, as all seed produced will arise from the desired cross.

Incompatability is a common phenomenon in cocoa which was discussed by Pound (1931) in his 'Studies of fruitfulness in cocoa', but it is only recently that an explanation of the mechanism, which is unlike that of most other plant species, has been put forward (Cope, 1958). The genetic explanation of incompatibility in cocoa need not concern us here, but its effect does. In the Trinitario populations first studied, self-incompatible trees are also cross-incompatible, and must be fertilised by pollen from a self-compatible tree. Where self-incompatible trees are common, this could have a depressing effect on yield, but, contrary to earlier ideas, a selection which is self-incompatible can yield heavily and several selections propagated widely (e.g. ICS 39, 40, 60, 89 and GS 29) are self-incompatible. Amazon selections are usually self-incompatible but are cross-compatible; Amelonado cocoa is self-compatible.

This outline of the incompatibility position is a simplified one and may not be entirely accurate, but the genetic mechanism of incompatibility, which continues to be unparalleled, is complex, resulting in a situation for which there is no simple description.

In Ghana many large seed gardens have been planted with the two parent clones vegetatively propagated. Their design has depended on whether one or both parents is self-incompatible. Where both are self-incompatible, all the pods will result from cross-pollination and can be used for seed, there being no apparent difference between a cross and its reciprocal. In such cases, equal numbers of each parent were planted, often in double rows of each clone. Where one clone is self-compatible, seed is gathered from the self-incompatible parent and the pollen parent is planted in the ratio one to five female parents.

Seed gardens must be isolated to some extent from other cocoa. A distance of 200 m (220 yd) is considered sufficient to prevent unwanted cross-pollination, but where some contamination is permissible because it will give acceptable seed, then it is simpler to surround the seed garden with several rows of the male parent, eight rows being the practice in Ghana.

These designs used in Ghana have not proved successful because of the difference in vigour of the two parents. In all cases one parent is an Amelonado or local Trinitario selection less vigorous than the other Amazon parent. Apart from difficulties in planting and establishment, the difference of vigour leads to suppression of the weaker clone, and this leads to weaker flowering and poor pod setting. There may also be different flowering periods affecting yield of seed pods.

Another form of seed garden has been described by Edwards (1969). In this, a series of self-incompatible parents are planted in such an order that a number of different crosses are produced, and seed can be collected from all the trees.

These two general types of seed garden were designed to depend on natural pollination. It is, of course, practical to plant a small number of plants of several clones and obtain seed of known crosses by hand pollination. It has been estimated that 0·4 hectare (one acre) of such a seed garden could produce enough seed to plant 160 ha (400 acres) in the fourth year rising to 480 ha (1,200 acres) by the eleventh year (Toxopeus, 1969).

Hand pollination coupled with a new design of seed garden can also overcome the difficulties which were encountered in Ghana. In future it is likely that seed gardens will comprise a large monoclone block of the female parent with a small area of the male parent alongside; hand pollination will be employed. Apart from overcoming the difficulties arising from differences of vigour, this arrangement will also allow the crop to be timed so that seed is available when it is needed, at the beginning of the wet season. It will also be possible to change the male parent, should this prove necessary.

Hand pollination

Hand pollination of cocoa is a fairly simple process but requires good eyesight, a steady hand and a pair of forceps. Pollen is collected by plucking freshly opened flowers from the male parent. The anthers must be removed carefully from these flowers and applied to the style and stigma of a female flower. The style is surrounded by a ring of staminodes and, when these are long, removal of two or three staminodes makes access to the style easier. Where flowers are plentiful a good operator should be able to do 600 to 800 pollinations per day and this should result in 250 to 300 pods. This rate of working includes the marking of pollinated flowers, which is generally sufficient for normal seed production. For breeding purposes it is necessary to cover the flowers with hoods of nylon mesh or small plastic tubes to prevent any natural pollination.

Movement of seed

Cocoa seed has no period of dormancy and the processes of germination will start soon after the beans are removed from the pod; beans can, in fact, germinate inside the pod in some circumstances.

The lack of dormancy does not cause any difficulty where the source of seed is close to the nursery, in which case the pods will be taken to the nursery and opened there. Seed can be kept in pods for seven to ten days with little loss in germination and pods can be moved a considerable distance in that time. There are, however, objections to moving pods between cocoa growing countries on the score of the possibility of introducing pests or diseases on the pod husk. Furthermore, the transport of pods involves carrying four times the weight of the beans themselves.

Early trials on the storage of cocoa seed tested the packing of seed in charcoal and showed that the seed could be preserved for as long as 13 weeks by packing the seed in charcoal powder with a moisture content of about 30 per cent and held in tins affording access of air (Evans, 1950). It must be emphasised that with this method the seed will germinate during storage so that the recipient will be faced with the problem of planting seed with shoots and radicles in the nursery.

When polythene became available and the advantage of such a moisture-proof barrier was appreciated, a new method was evolved by Alvim (1958). There have been some modifications to the original method and the method now recommended is as follows:

The beans are mixed with sawdust and the testas are removed; the peeled seeds are treated with a fungicide either by washing in a fungicidal solution or by dusting. The treated seeds are packed in polythene bags, each bag holding about 500 seeds or about 1 kg. Seed stored by this method will preserve its viability without germinating for three to four weeks. Experiments have given the following results (Hunter, 1959):

No. days after removal from pods	Germination (%)
0	96
14	86
21	78
28	70

The task of removing the testas may seem unnecessary, difficult and possibly expensive. It does, however, make the application of fungicide relatively easy and it has been found that a practised person can peel 200 seeds per hour. If 500 seeds are needed for 0·4 ha (1 acre) then the cost of peeling is a relatively small additional cost.

The choice of fungicide is of some importance. Hensen and Hunter (1960) listed several fungicides and all except Captan helped to prolong viability. Swarbrick (1965) working in Nigeria, tested six fungicides; three gave over 80 per cent germination after five weeks, one failed to preserve viability and two gave intermediate results. Ziram was one of the effective fungicides; the failure was Thiram which had proved effective in Costa Rica.

References

Alvim, P. de T. (1958) 'Un procedimiento simple para conservar el poder germinativo de las semillas de cacao', *Septima Conferencia Interamericana de Cacao, Colombia 1958*, pp. 277–82.

Anon (1970) *The CRIG Series II Hybrids. The position at March 1970*, Memorandum Cocoa Res. Inst. Tafo.

Atanda, C. A. and Toxopeus, H. (1971) 'A proved case of heterosis in *Theobroma cacao* L.', *III Internat. Cocoa Res. Conf. Accra 1969*, pp. 545–51.

Bartley, B. G. D. (1969) 'Twenty years of cacao breeding at the Imperial College of Tropical Agriculture, Trinidad', *Proc. 2nd Internat. Cacao Res. Conf. Bahia 1967*, pp. 29–33.

Bridgland, L. A. (1960) 'Cacao improvement programme, Keravat', *Papua and New Guinea Agric. J.*, **12**, 149–67.

Chee Yau Fah (1970) 'Cocoa breeding in Sabah', *Cocoa Seminar Tawau 1970*, Sabah Planters Assn., pp. 35–43.

Cope, F. W. (1958) 'Incompatibility in *Theobroma cacao*', *Ann. Rep. Cacao Res. 1957–58*, Trinidad, pp. 7–17.

Edwards, D. F. (1969) 'Hybrid seed gardens: some practical considerations', *Cocoa Growers' Bull.*, **13**, 14–19.

Evans, H. (1950) 'Results of some experiments on the preservation of cocoa seed in viable condition', *Trop. Agric.*, **27**, 48–55.

Freeman, W. G. (1929) 'Cacao research. Results of cacao research at River Estate, Trinidad', *Trop. Agric.*, **6**, 127–33.

Freeman, W. E. (in press) 'Cocoa breeding', *Proc. 4th International Cocoa Res. Conf. Trinidad 1972*.

Glendinning, D. R. (1964) '*Phytophthora* cankers and black-pod disease', *Rep. Period 1 April 1962–30 September 1963, Cocoa Res. Inst. Tafo*, p. 52.

Haddon, A. V. (1961) 'Variety trials of seedling cacao in Malaya', *Malayan Agric. J.*, **43**, 169–232.

Hansen, A. J. and Hunter, J. R. (1960) 'A preliminary experiment on the protection of cacao seeds', *Proc. 8th Inter-Amer. Cacao Conf. Trinidad 1960*, pp. 121–4.

Hunter, J. R. (1959) 'Germination in *Theobroma cacao*', *Cacao*, **4** (4), 1–8.

Jolly, A. L. (1956) 'Clonal cuttings and seedlings of cocoa', *Trop. Agric.*, **33**, 233–7.

Jolly, A. L. and de Verteuil, L. L. (1960) 'Clonal and fertiliser experiments in Trinidad and Tobago', *Proc. 8th Inter-Amer. Cacao Conf. Trinidad 1960*, pp. 390–418.

Lodos, N. (1967) 'Studies on *Bathycoelia thalassina* (H-S) (Hemoptera, Pentatomidae), the cause of premature ripening of cocoa pods in Ghana', *Bull. Ent. Res.*, **57**. 289–99.

Posnette, A. F. (1951) 'Progeny trials with cacao in the Gold Coast', *Emp. J. exp. Agric.* **19** (76), 242–52.

Pound, F. J. (1931) 'Studies in fruitfulness in Cacao II—Evidence for partial sterility', *1st Ann. Rep. Cacao Res. 1931*, Trinidad, pp. 24–5.

Pound, F. J. (1932) 'The genetic constitution of the cacao crop', *2nd Ann. Rep. Cacao Res. 1932*, Trinidad, pp. 9–25.

Pound, F. J. (1938) *Cacao and Witchbroom disease of South America with notes on other species of* Theobroma, Trinidad.

Pound, F. J. (1943) *Cacao and witches' broom disease* (Marasmius perniciosus); *report on a recent visit to the Amazon Territory of Peru*, Trinidad.

Russell, T. A. (1952) 'The vigour of some cacao hybrids', *Trop. Agric.*, **29**, 102–6.

Shepherd, C. Y. (1937) *The Cacao Industry of Trinidad. Some economic aspects*, Series III. An examination of the effects of soil types and age on yields. Trinidad.

Swarbrick, J. T. (1965) 'Storage of cocoa seeds', *Expl. Agric.*, **1**, 201–7.

Toxopeus, H. (1964) 'F3 Amazon cocoa in Nigeria', *Ann. Rep. West Afr. Cocoa Res. Inst., Nigeria 1963–64*, pp. 13–23.

Toxopeus, H. (1969) 'The use of small polyclonal seed orchards on commercial cocoa plantations', *Cocoa Growers' Bull.*, **12**, 14–16.

Toxopeus, H. (1972) 'Cocoa breeding: a consequence of mating system, heterosis and population structure', *Cocoa and coconuts in Malaysia, Proc. Conf. Incorp. Soc. Planters, Kuala Lumpur 1971*, pp. 3–12.

Vernon, A. J. (1971) 'Canker: the forgotten disease of cocoa', *Cocoa Growers' Bull.*, **16**, 9–14.

Wanner, J. A. (1962) *The First Cocoa Trees in Ghana, 1858–1868*, Basle.

Chapter 6

Nurseries and Vegetative Propagation

Choice of seedlings or rooted cuttings

Cocoa is normally planted as seedlings but there are occasions when some form of vegetative propagation is preferable. Seedlings are normally used because there is rarely a shortage of seed, they are cheap to produce and growth of seedlings gives trees with a convenient habit of growth. On the other hand, rooted cuttings—the usual method of vegetative propagation with cocoa—have several disadvantages: material is often limited in quantity, they are costly to produce and, in the early years at least, the plants are bushy which makes field operations more difficult. These difficulties could be overcome by the use of budding, the techniques of which have been established for many years, but which have not been used extensively partly because of the danger of chupons from the stock replacing scion growth and partly because of the difficulties involved in the supervision of large numbers of budded plants in the early stages.

The use of seedlings is of doubtful value where the planting material does not breed true and improvement cannot be effected by using seed from selected trees. This is particularly true of Trinitario cocoa. The selection of high-yielding trees and the development of reliable methods of raising rooted cuttings during the 1930s led to the use of rooted cuttings on a wide scale in Trinidad and in many other countries in the Caribbean area. The use of rooted cuttings was largely successful in giving a more uniform stand of trees but the method was costly and has been replaced by the use of seed of known parentage and performance. In other countries the original use of unselected seed was superseded by the provision of seed from trees of known parentage and later by hybrid seed from seed gardens in which the two parents are planted as rooted cuttings.

Cocoa seed is non-dormant and must be sown shortly after the pods have been harvested. The measures taken to preserve seed viability for the purpose of transport have been described, but seed from any source must be sown without delay.

Sowing seed at stake

Cocoa seedlings may be raised in a nursery, or seed may be sown at stake. The vast areas of cocoa in West Africa have largely been established by sowing at stake, the farmers sowing three seeds in a group, placing each just under the surface of the soil. While this method can be successful and obviously requires little labour, it has several important disadvantages. In the first place, it is extravagant in its use of seed as it requires two or three times the normal quantity. This factor alone would rule it out where seed is in short supply or has had to be transported over a long distance. Apart from this aspect, seed at stake often suffers severely from rodent attack. In Ghana, seed at stake sown in May–July at the beginning of the wet season is badly attacked (Hammond, 1962); sowings in September–October suffer less but as the rains end in this period establishment is more hazardous. In Papua–New Guinea, different opinions have been expressed: Green (1938) thought that planting at stake should not be used for several reasons, including the risk of rodent damage; more recently Henderson (1954) has recommended planting at stake whenever possible. A recent trial in Malaya comparing seed at stake with seedlings showed that there were greater losses with seed at stake despite the sowing of two seeds at each point; heavy losses occurred when very heavy rainfall occurred after sowing (Teoh and Shepherd, 1972).

Another disadvantage of seed at stake is that it makes the selection of more vigorous seedlings more difficult and the method does not allow any flexibility in planting time which might be necessary when planting conditions are unsuitable. Apart from these disadvantages, nursery seedlings are likely to develop more rapidly.

A trial in Nigeria compared seedlings sown in a nursery during October–December and planted out during the following May–July, with seed sown at stake during the latter three months; the germination of the seed at stake was uneven and lower than normal and after two dry seasons the percentage of seedlings that had jorquetted was much lower with the seed at stake than with seedlings raised in a nursery (Freeman, 1964).

Nursery practice

A cocoa nursery will require shade, water and protection from wind; it may also be necessary to provide protection from rodents. Shade can be provided but in choosing a site the supply of water and position in relation to the area to be planted must be considered.

Most nurseries are used for only a few years so that a simple structure of posts and cross pieces will suffice to provide a framework for shade and lateral protection. Palm fronds are often used for this purpose as the shade can easily be adjusted and thus have the advantage of letting in more light as they decay. The initial shade is usually quite heavy, somewhat in excess of 50 per cent, but decreases as the seedlings grow; when the plants are ready for planting out, the shade should be of the same degree as the shade in the field. Cocoa seed is sometimes raised in a seedbed and subsequently transferred to a basket for further growth in the nursery. This practice has nothing to commend it. It is preferable to plant the seed straight into a nursery container. Cocoa seed has a high rate of germination so economies in baskets from the use of an initial seedbed would be negligible and the act of transplanting from seedbed to basket is liable to damage the root system and cause a check to growth.

Containers for seedlings have been made of a great variety of materials; woven split cane, bamboo, veneer tubes and proprietary materials, but all these have been superseded by polythene bags, which are usually cheaper, more durable and simpler to store and transport. The bags are made of light gauge polythene—350 gauge, or 0·08 mm— and are usually supplied with one or more drainage holes.

The size of the pot or bag is of some importance as one that is too small will restrict root growth while too large a bag will require more soil, take longer to fill and take up more space. In many cases the seedlings will be in the nursery for up to five months and for this period of time the bags should be 25 cm long and 10 to 12 cm in diameter when filled. Polythene bags are usually gusseted and specified on their layflat size. In Malaya polythene bags 30 by 20 cm layflat are used where the period in the nursery is four to five months and 23 by 18 cm or 25 by 18 cm where the period is two and a half months (Leach *et al.,* 1971).

In West Africa top soil alone is used for filling the bags; the top soil is generally well sieved to remove roots but manures or composts are not used to any extent. Nursery practices have not been subjected to extensive trial but one trial of the effect of potting soil on germination and growth showed that germination was slightly better on heavy soils but subsequent growth was better on lighter soils (Wessel, 1966). In Malaya where heavier soils occur mixtures of top soil and coarse river sand in the ratio of 2:1 are used on some plantations and Whitehead (1954) found that the following mixture gave fairly uniform growth: seven parts loam, three parts dried farmyard manure and two parts sharp sand.

The bags are arranged in beds which can be any convenient length but are normally only 0·6 to 1 m wide. This allows easy inspection and watering of all the plants. Support for the bags is usually provided by strips of bamboo.

Cocoa seed is epigeal in its growth, i.e. the cotyledons are raised above the soil surface by growth of the root. It is therefore necessary to ensure that the seed is not planted too deep else it may fail to emerge.

Seed is planted no more than 1 cm below the surface with the hilum downwards but it is safer to lay the seed flat rather than to plant on end. This will avoid any possibility of planting the seed with the hilum upwards in which case it is possible under some conditions for highly distorted seedlings to develop.

Cocoa beans germinate rapidly and germination of a batch of seed will usually be complete within two weeks and the rate of germination should be at least 90 per cent.

The beans are not normally treated in any way before sowing. The mucilage decomposes within a short space of time; washing the beans or rubbing them in sawdust or sand to remove the mucilage will lead to germination taking place a day or two earlier but the advantage is marginal. Occasionally insects are attracted by the mucilage and where this takes place the beans should be washed. Seed dressings with insecticides can be used but are rarely necessary; cocoa seed, like many others, is affected by BHC which will delay and reduce germination if a heavy overdose is applied; normal doses can be tolerated (Brown, 1968).

The nursery will require watering, adjusted according to local conditions. In Nigeria an experiment compared watering every other day with watering twice a week; the less frequent watering gave the same results under shaded conditions. The usual error is overwatering, which can lead to fungal attack on the stem of the seedlings or rot of the young tap root. Apart from watering, the seedlings need little attention while in the nursery. There may be attacks by rodents or by insects but these can normally be countered by physical protection or by use of insecticides. Some insects, such as crickets, can be controlled by application of DDT or BHC powder around the boundary of the nursery or around each bed. Other flying insects may require application of insecticides to the seedling.

The application of fertilisers in the nursery has been found to be of no benefit where normal top soil is used, but a mixture of fertilisers and organic manures can make up for the deficiencies of a poor potting medium (Wessel, 1969).

Containers in which the beans have failed to germinate can be removed from the beds and re-used.

Seedlings which are smaller than average should either be moved to another bed to prevent their being crowded out or should be thrown away if they are weak.

The length of time that the seedlings remain in the nursery may be dictated by the time when seed is available and the time when planting can be done; in West Africa for instance seed is plentiful in October–December but planting must await the rains at the end of March or early April. In these circumstances the seedlings will be kept in the nursery for four to five months at which age the plants will be 40 to 50 cm tall. Plants should not be kept longer than six months in the nursery else they will be liable to suffer a setback at planting from which they will not completely recover.

In the trial in Malaya mentioned earlier seedlings aged one to six months were planted out at different times of the year (Teoh and Shepherd, 1972). In wet months losses varied inversely with age at planting, but in a drier month losses did not vary with age at planting. Highest initial production was recorded from four-month-old seedlings, and this seemed to be the maximum age under Malayan conditions. Where growth is good and planting conditions suitable the seedlings can be planted out earlier, but it is unusual to plant out seedlings which are less than three months old.

Vegetative propagation

The need for vegetative propagation arises when selected trees will not breed true; this was recognised many years ago in Trinidad and applies to Trinitario cocoa generally because its progeny are highly variable in performance. Methods of budding and grafting were known before 1900 but were little used because reliable methods of selection had not been evolved. Such methods were formulated in the early 1930s in Trinidad and at the same time the rooting of cuttings was investigated. Budding and grafting were not employed, mainly because the stocks would have been highly variable but also because there was a risk of stock–scion interaction and the possibility of the stock throwing up a chupon which might take over.

Rooted cuttings

The techniques of rooting cuttings were developed initially by Pyke (1933) but were re-examined in detail by Evans (1951), who clarified the conditions for successful rooting. The early methods using closed bins were used successfully on a large scale in Trinidad, Grenada and other islands in the Caribbean, and to a lesser extent in Central and South America. This method offered better planting material but suffered from a number of disadvantages; it was costly not only in the capital investment but also in running costs. Various other methods, such as spray beds and humidified glasshouses, were evolved but required greater skill and special conditions. With the development of polythene sheet, simpler, cheaper methods became possible and these can be used on any scale.

The plants produced from rooted cuttings have a low, spreading habit of growth which makes field work and harvesting more difficult during the early years. These difficulties can be reduced by pruning as the tree grows but the seedling habit of growth is easier to manage. The development of seed of proven ability has led to a sharp decline in the

18. Propagating bins, Trinidad.

19. (*right*) Rooted cuttings developing prior to planting.

20. (*above left*) Young cocoa planted under thinned jungle, Cameroon.

21. (*left*) Land cleared for cocoa planting, New Guinea. The area on the right is planted with *Leucaena leucocephala*.

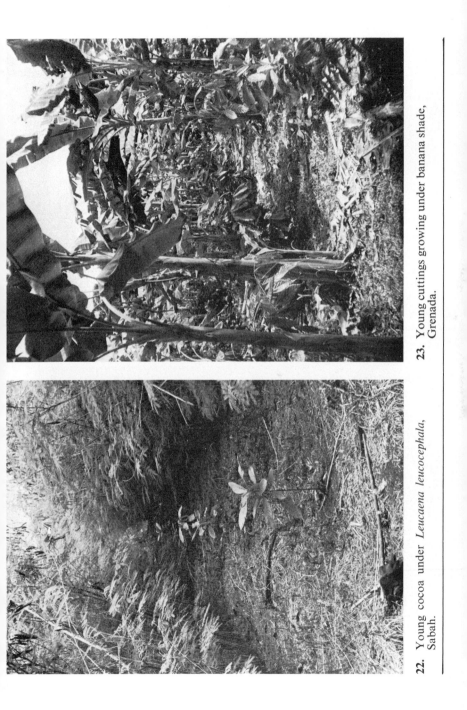

22. Young cocoa under *Leucaena leucocephala*, Sabah.

23. Young cuttings growing under banana shade, Grenada.

24. Cocoa under *Leucaena leucocephala*, Papua—New Guinea

25. Cocoa under *Terminalia ivorensis*, Tafo, Ghana

26. (*left*) Cocoa shaded by *Albizzia chinensis*, Sabah.

27. Cocoa under *Gliricidia maculata* and Dadap, West Malaysia.

28. Cocoa shaded by *Erythrina velutina*, Mexico.

29. Criollo cocoa under *Erythrina lithosperma* (Dadap), India.

30. Coconuts interplanted with cocoa, West Malaysia.

31. (*right*) Cocoa planted under irrigated arecanut, India. The shade is too heavy for optimum growth and yield.

32. *(left)* Cocoa in Ghana showing typical mixed jungle shade.

33. Cocoa in Bahia, Brazil.

use of rooted cuttings for commercial plantings in the West Indies. This decline will probably continue but vegetative propagation still has an important function to play in the testing of selections and in seed gardens.

Conditions required for rooting cuttings

The successful rooting of cocoa cuttings depends on a large number of factors, but the most important are the conditions of temperature, light and humidity in which the cuttings are kept. The various factors affecting the process of rooting are in order of sequence:

1. *Type of cocoa* Wide variations have been found in the rooting ability of selections. While many of the ICS clones are easy to root, others are not, and the 'Nacional' selections from Ecuador are particularly difficult. These differences are nutritional and may be overcome by injection with nutrients and by treatment with hormones.

2. *Nursery management* For large-scale production of cuttings a special nursery is planted at a close spacing of 2×2 m. The nursery is shaded with bananas and Gliricidia to allow 20 to 50 per cent of full sunlight on to the nursery plants. Cuttings can be taken from the plants a year after planting and each plant should produce twenty-five cuttings a year. Thus 0·4 hectare (one acre) of nursery at 2×2 m will produce 30,000 cuttings, sufficient for planting about 16 hectares (40 acres) at 3×3 m, assuming normal losses. Nursery plants need fertilising regularly and in Trinidad the application of sulphate of ammonia and a complete fertiliser alternately at a rate of 80 to 115 g (3 to 4 oz) per plant every ten to twelve weeks is recommended. As the trees age the cuttings become less vigorous and after six years the nursery will have to be replaced.

3. *Age of flushes* For successful rooting flushes should be in the process of hardening when taken for cuttings. This semi-hardwood stage is indicated by browning of the upper surface of the shoot.

4. *Use of hormones* Trials with a variety of substances have shown that the optimum treatment is to dip the cut end in a solution of 8 to 10 g of equal parts of α-naphthalene-acetic acid and β-indole-butyric acid in a litre of 50 per cent alcohol (Evans, 1951).

5. *Temperature, light and humidity* The function of light is to provide carbohydrates through photosynthesis. If light is inadequate the cuttings will starve and fail to root; this is indicated by yellow areas which first appear at the base of the leaves. If light is excessive carbohydrates are produced in excess of the supply of nitrogen and the leaves turn pale

yellow and are liable to break down.

Temperature also affects rooting. As temperature rises so will the rate of respiration and demands for carbohydrate. It becomes more difficult to maintain high humidity as the temperature increases and humidity must be maintained at near saturation as the cutting has initially no means of replacing water losses. To provide these conditions cuttings are heavily shaded and kept cool and moist by water sprays during the rooting period.

6. *Rooting medium*   For quick successful rooting it is vital that the rooting medium should provide the optimum conditions of air and moisture round the base of the cutting. If the medium is waterlogged the cutting will rot; if oxygen supply is inadequate, which can result from overwatering, hard callus rods will be produced and rooting considerably delayed; if conditions are not wet enough a pad of hard callus tissue may develop and rooting prevented or delayed. Optimum conditions will lead to growth of a prolific root system within fourteen days but twenty-one days is the usual duration in practice.

The closed bin method

Several methods have been evolved which provide these conditions under a variety of circumstances. The first method used on a large scale involved the use of closed bins. These bins are constructed of bricks or concrete blocks and covered by glass. The propagating unit is shaded artificially and the light in the bins reduced further by covering the glass with newspaper or cloth, but care must be exercised in selecting the cloth as the amount of light transmitted varies widely. It is desirable that the cloth should transmit 650–750 foot candles in full sunlight. Temperature is kept down by keeping the paper or cloth wet. Stones and gravel are placed in the bottom of the bins to provide drainage beneath the layer of rooting medium. Various substances have been used for rooting, the most usual being composted sawdust, coconut fibre dust or sand.

The stem cuttings with four or five leaves are taken from the nursery trees early in the morning and treated by trimming the leaves to about half their normal size and dipping them in hormone solution. In taking cuttings it is essential to leave one or two buds on the flush. The treated cuttings are placed in the rooting bins, with the cut ends about 6 cm below the surface and the cuttings spaced so that they do not shade one another. The rooted cuttings are removed from the rooting bin after four weeks and potted in potting mixture. As the root system is not capable of meeting the needs of the plant at that stage the potted cuttings are placed in hardening bins for a further two weeks. During that period the time of exposure to ambient conditions is increased by stages.

This briefly describes the method used for several years by the Cocoa Board in Trinidad which produced about 1·5 million plants a year (Moll, 1956a). The rooting percentage achieved was 60 to 70 per cent but, with further losses during hardening and storage, the production efficiency—the proportion of usable plants produced from 100 cuttings —was normally 50 per cent.

Modifications and improvements

There have been several modifications to this basic technique including open-spray beds and humidified glasshouses. In an open-spray bed the atmosphere around the cutting is kept saturated by means of a constant fine spray; this method requires a good supply of pure water at a pressure of 1 kg/cm^2 (50 lb/in^2). The humidified glasshouse requires a centrifugal humidifier to maintain a saturated atmosphere. This equipment needs less water than an open-spray bed, but a supply of electricity is essential. These methods have had limited use.

The major modification to the standard method has been the pot rooting technique (Murray, 1954). This involves the use of baskets or pots containing potting mixture with a central core of rooting medium, so that the cutting can be rooted and hardened without transplanting. The advantages of this method lie in reduction in labour and an increase in production efficiency which, in Trinidad, was raised from 45 per cent to 65 per cent when this technique was adopted (Moll, 1956b).

Another development is the polythene sheet method evolved in Ghana (McKelvie, 1957). In this method cuttings are planted in a cored basket in the same way as in the pot rooting technique; a number of baskets are grouped together in a shaded nursery and covered with a polythene sheet which is weighed down at the edges. The baskets are watered thoroughly when the cuttings are inserted but a light watering every third day is all that is necessary thereafter. Two-leaf cuttings were used in the trials and shade in the nursery allows 15 per cent light on to the polythene sheet. Under these conditions roots appear through the baskets in four weeks and the plants are then hardened by removing the polythene sheet for a gradually increasing period. A high rate of success has been achieved with this method which is cheaper than the standard method and is far easier to use on a small scale.

Other methods of vegetative propagation

Apart from rooted cuttings cocoa can be propagated vegetatively by budding, grafting or marcotting.

Budwood has the advantage of being easy to transport over long distances and providing more material from a given source, so that budwood is the usual form in which selected trees are transferred from one country to another. The technique of budding is a skilled procedure and the budded tree has the disadvantage of being liable to produce a chupon from the stock. These disadvantages are unimportant as far as the above purpose is concerned, but they have discouraged the use of budding on a large scale. The yield of buddings has been compared with cuttings in some trials in Trinidad and it was shown that good clones do better as cuttings while poor ones may perform better when budded; presumably this is due to differing root systems (Cope, 1953). There is no experimental evidence to show that buddings have any long-term advantage in the field, and, while a budded plant may be cheaper to produce than a rooted cutting, the disadvantages prevent this method being used as a general means of propagation. Budwood can be taken from a tree without any special preparation but the tree should be at the stage when the leaves have just fallen. Budwood from chupons will develop as chupons and budwood from fans as fan growth. Normally the budwood should be used soon after collection but its viability can be preserved for several days by storing it in damp charcoal, moist sand or sawdust.

Several methods of budding have been used, the oldest method being a patch bud method described and illustrated by Van Hall (1932). A modification to this method was evolved by Bowman with his inverted U technique. This involves making three cuts and inserting the bud beneath the flap of bark; this method has been described by Rosenquist (1952). Topper (1958) developed an inverted T method, a form of shield budding, which involves the use of buds up to 4 cm long taken from mature terminal shoots. These buds are budded on to seedlings about four months old and are covered with clear budding tape, which is removed after three weeks.

The buds are cut from the stem, the shape varying according to the method employed, and fitted on to the seedling stock so that the cambial tissues of both stock and scion are in firm contact. The bud must be kept in place with raffia, waxed tape or transparent plastic tape which will prevent any loss of moisture. Firm wrapping and complete covering of the wound should ensure a high rate of success. The tape is removed after two to three weeks. At this stage buds may remain dormant for a long period but this can be overcome by removing a section of bark just wider than the original wound and 8 mm above the bud. This has been found to stimulate the buds (Van de Burg, 1969).

Cocoa has been grafted by saddle and wedge graft methods but they have little practical application. To detect virus infection in budwood it is necessary to graft the budwood on to a susceptible stock for which a rapid and reliable method is needed; a side-graft technique has been evolved for this purpose (Soderholm and Shaw, 1965).

Where it is necessary to propagate from older vegetative material, marcotting can be used. This involves removal of the bark in a strip 7·5 cm wide and covering the xylem with sawdust held in place by a strip of polythene sheet. The branch is cut off after roots have been formed. Okoloko (1965) achieved good results with this method but its use is limited to certain special situations.

References

Brown, D. A. Ll. (1968) 'Seedling development inhibited by B.H.C.', *Cocoa Growers' Bull.*, **11**, 27.

Cope, F. W. (1953) 'Some results of the clonal trials at River Estate', *Rep. on Cacao Res. 1945–51*, Trinidad, pp. 12–23.

Evans, H. (1951) 'Investigations on the propagation of cacao', *Trop. Agric.*, **28**, 147–203.

Freeman, G. H. (1964) 'Nursery and establishment trials', *Ann. Rep. W. Afr. Cocoa Res. Inst. (Nigeria) 1962–63*, pp. 61–5.

Green, E. C. D. (1938) 'Cacao cultivation and its application to the mandated territory of New Guinea', *Papua and New Guinea Agric. Gaz.*, **4** (4), 1–63.

Hammond, P. S. (1962) 'Cocoa Agronomy', in J. B. Wills, ed., *Agriculture and Land use in Ghana*, Oxford University Press, pp. 252–6.

Henderson, F. C. (1954) 'Cacao as a crop for the owner-manager in Papua and New Guinea', *Papua and New Guinea Agric. J.*, **9**, 45–74.

Leach, J. R., Shepherd, R. and Turner, P. D. (1971) 'Underplanting coconuts with cacao in Malaya', *Proc. 3rd Internat. Cocoa Res. Conf. Accra 1969*, pp. 346–55.

McKelvie, A. D. (1957) 'The polythene sheet method of rooting cacao cuttings', *Trop. Agric.*, **34**, 260–5.

Moll, E. R. (1956a) *Instructions for propagation of cocoa*, Cocoa Board Circular No. 1 of 1956. Trinidad.

Moll, E. R. (1956b) 'The pot rooting technique of cacao propagation', *Proc. 6th Inter-Amer. Cocoa Conf. Bahia 1956*, pp. 221–7.

Murray, D. B. (1954) 'A new method of vegetative propagation', *Proc. 5th Inter-Amer. Cocoa Conf. Turrialba 1954*, Doc. 7.

Okoloko, G. E. (1965) *Cacao Marcotting*, Memorandum No. 8 Cocoa Res. Int., Nigeria.

Pyke, E. E. (1933) 'The vegetative propagation of cacao, II. Softwood cuttings', *2nd Ann. Rep. on Cacao Res. 1932*, Trinidad, pp. 3–9.

Rosenquist, E. A. (1952) 'Notes on the budding of cacao', *Malayan Agric. J.*, **35**, 78–84.

Soderholm, P. K. and Shaw, E. W. (1965) 'A modified side graft technique for use in a cacao virus indexing program', *Proc. Amer. Soc. Hort. Sci. Caribbean Reg.*, **9**, 25–9.

Teoh, C. H. and Shepherd, R. (1972) 'Age at planting of cocoa seedlings in relation to month of planting', *Cocoa and coconuts in Malaysia. Proc. Conf. Incorp. Soc. Planters, Kuala Lumpur 1971*, pp. 76–85.

Topper, B. F. (1958) 'A new method of vegetative propagation for cocoa', *Rep. Cocoa Conf., London 1957*, pp. 49–51.

Van De Burg, B. (1969) 'Cacao budding—a neglected technique', *World Crops*, **21**, 105–7.

Van Hall, C. J. J. (1932) *Cacao*, 2nd ed., Macmillan.

Wessel, M. (1966) 'Effect of potting soil on germination and growth of cocoa seedlings', *Ann. Rep. Cocoa Res. Inst. Nigeria, 1964–65*, pp. 87–8.

Wessel, M. (1969) 'Potting medium for cacao seedlings', *Ann. Rep. Cocoa Res. Inst., Nigeria, 1967–68*, pp. 38–9.

Whitehead, C. (1954) 'Cacao propagation in Malaya with special reference to cacao nursery techniques', *Malayan Agric. J.*, **37**, 203–10.

Chapter 7

Establishment

There are two general methods by which cocoa is established: by planting permanent shade following clear felling, and by thinning jungle. In addition, cocoa can be interplanted between coconuts and some other crops where suitable conditions prevail.

Clear felling and planting of shade was commonly practised in the West Indies and South America and has been used in some of the cocoa-growing countries in the Far East. The method involves clear-felling the jungle, followed usually by burning the felled timber, planting shade both permanent and temporary, and finally planting cocoa. The main advantages of this method lie in the regular pattern of shade it provides and the ease of planting and maintenance that results; on the other hand, the method is expensive and very often leads to a longer period for establishment.

Planting under thinned jungle is employed almost exclusively throughout West Africa and has been used in some other countries. The method is cheap and simple—well suited to the small farmer—and relatively quick, but it has disadvantages: the shade is uneven and difficult to regulate; some jungle trees have proved to be competitive and others are alternate hosts to virus diseases and pests. Before dealing with these methods in detail the conditions required at the time of planting and during the establishment period must be described as a grower must have these clearly in mind when deciding on his particular method and timing of operations.

Requirements for shade

The general question of shade and fertilisers as applied to mature cocoa is described in another chapter, but, to summarise the results of recent experiments, shade can be dispensed with where the trees can obtain adequate nutrients and moisture throughout the year. Under these conditions high yields can be expected, but such conditions are unusual and in most countries some shade is needed for mature cocoa.

For young cocoa shade is always recommended. One of the main reasons is to ensure the right form of growth. The amount of light falling on a young tree will influence the way it grows, low light intensities or heavy shade leading to long internodes and few side branches, high light intensities or little shade giving the opposite effect, which leads to bushy growth. Too much light is therefore undesirable as it will delay the time when, at normal spacings, a canopy will be formed, and the early formation of a canopy is necessary to reduce weed growth and to allow shade to be reduced in order to induce flowering and cropping.

It has been stated that it is possible to bring cocoa into bearing without any permanent shade. This was based on trials in Nigeria where the dry season is often quite severe but Freeman (1964) concluded that establishment without shade required 'the best establishment methods' and 'could not be recommended as a sound commercial practice'. The 'best' methods used in the trials involved close spacing at 1.2×1.2 m (4×4 ft) with the intention of subsequent thinning; it was also necessary to control pests which attack unshaded cocoa. This method of establishment might be easier where the dry season is less severe but special measures would be needed including close spacing initially.

It is recommended therefore that young cocoa should be shaded and also protected from wind. The question is, how much shade? The shade and fertiliser experiments conducted in Trinidad showed that 50 per cent shade gave the greatest early growth and highest initial yields (Murray, 1953). These results apply to Trinidad conditions where hours of sunlight are long and light intensity is high; 50 per cent shade can mean something quite different in terms of incident energy in other climates. Where light intensity is lower less shade would seem to be required. However, data on light intensity or incident energy have not been gathered in many places·so it is difficult to make comparisons between cocoa growing countries. The usual recommendation is, therefore, 50 per cent shade for young cocoa, and, as a rough guide, this is sufficient.

It should be stressed that suitable shade conditions must be provided by the time the cocoa is due to be planted. Where land is clear-felled and shade established there is a temptation to plant cocoa as soon as the season is right but long before the shade is adequate. This will result in a severe shock to the cocoa plant which has probably been raised in a heavily shaded nursery; this violent change of conditions can be minimised by providing artificial shade for the young plants, but the added expense may be considerable. The alternative course of action— to delay planting until shade conditions are suitable—is to be recommended.

Planting after clear-felling

The clearing of jungle has to be done during the dry season and is usually initiated by clearing the undergrowth at the end of the wet season or as soon as the dry season starts. This initial operation allows easier access to the land for the subsequent felling and it also makes it easier to assess the slopes and topography which may affect the layout of roads and any drains that may be needed. The need for drains depends on the rainfall, topography and type of soil and where they are needed it is easier to dig them before the forest is felled.

After these initial operations the forest is felled and the trees cut up to make clearing the land easier; where the trees and rubbish are to be burnt, the trees must be cut up to allow them to be stacked. All this must be done well before the end of the dry season, otherwise it will be impossible to burn the trunks completely. An inadequate burn results in the need to restack and reburn, an additional expense. There are countries where the seasons are not sufficiently marked for these operations to be conducted easily; in such cases drainage may help to provide drier ground conditions.

The practice of burning after clear-felling is criticised on the grounds that it destroys a great deal of potential humus, and releases large amounts of ash, often locally, which may cause imbalance of nutrients; it also exposes the soil to sun and rain which can lead to severe erosion and more certainly to losses of organic matter in the surface layer. The advantages of burning after clear-felling, as compared with leaving the fallen timber to disintegrate, lie in the ease of later operations and their supervision; the presence of large amounts of timber on the ground following clear-felling of thick jungle makes lining, holing, planting and any further operations extremely difficult and these difficulties are likely to endure for several years. To most planters the advantages resulting from burning outweigh any losses and dangers that result from it, and awareness of the dangers arising from erosion and the loss of organic matter should result in steps being taken to minimise them.

The dangers of clear-felling do not arise solely from the resulting exposure of the soil but also from the disturbance of the soil during the operation. Where land is cleared by hand disturbance is minimal, but where heavy machines are used for felling, stumping and wind-rowing, the soil may be considerably disturbed and compacted. This can easily damage the structure of the soil, especially heavier soils, and this damage will be increased if the organic matter is subsequently destroyed by prolonged exposure. On deep alluvial soils, heavy machines may be used without causing much harm but their use on soils with a shallow humic layer is likely to be harmful.

Methods of mechanical clearing of forest have been developed for the planting of oil palms and rubber which allow large areas to be developed in a relatively short time. After clearing the land, cover crops are planted followed by the palms or rubber at a planting density of 125 to 250 trees per hectare. With cocoa, shade has to be planted after clearing the land and a much greater planting density is used. This slows down a planting programme, as hand labour is involved. This means that planting programmes for cocoa are much smaller than for oil palms and rubber, so that mechanical clearing may not be a practical proposition.

Planting shade

The next stage is for the area to be lined and shade to be planted. The question of shade for young cocoa is complicated by the changing needs of the cocoa tree as it matures. Heavy shade will be needed initially but this must be readily adjusted in the first few years, leaving in the end a small number of trees as shade for the mature cocoa. These changes infer that a mixture of shade trees and sometimes cover crops will be needed in the first place. The properties required of an ideal shade tree have been described as follows (Freeman, 1964):

> It will be easy to establish and provide a good shade throughout the dry season yet not compete excessively with the cocoa roots for moisture and soil nutrients. It will be easy to remove when finished with, yet its removal will not damage the cocoa canopy. Further, it should not be an alternative host species to insect pests of cocoa. Finally, if possible, it will be of commercial value.

It is difficult, if not impossible, to find all these properties in one shade tree, and a species that comes near to this ideal in one country may be quite unsuitable in another. While one or two species may provide permanent shade, additional shade will be needed during the initial establishment stage and this may best be provided by other quick-growing shade or cover crops. The aim should be a mixture of species which will rapidly provide the right degree of shade for cocoa seedlings and then be easily adjusted as both shade and cocoa develop.

Permanent shade trees

The trees which have been commonly used as permanent shade and the countries where they have been planted are:

Trinidad	*Erythrina poeppigiana* Mountain Immortelle
	Erythrina glauca Swamp Immortelle
Papua–New Guinea	*Leucaena leucocephala*
Sri Lanka	*Erythrina lithosperma* Dadap
Java	*Ceiba pentandra* Kapok
Mexico	*Erythrina velutina*

These and many other species are used as permanent shade and the following notes describe the main features of some of them:

Leucaena leucocephala This small leguminous tree grows quickly and provides light feathery shade. It is easily grown from seed and, in Sabah, has been sown in hedges using up to 28 kg seed per hectare (25 lb per acre); experience in New Guinea has shown that it is preferable to sow the seed in clumps at 60 cm (24 in) intervals giving three clumps of Leucaena for each cocoa tree rather than to sow in a continuous line (Newton, 1966). In New Guinea it has been used extensively as permanent shade and was for many years favoured because it was also free from pests and diseases. In recent years *L. leucocephala* has been heavily attacked by defoliating caterpillars which readily transfer their attention to the cocoa tree beneath (Dun, 1967). This type of attack is much less common under coconuts and bush shade, so *L. leucocephala* has become less popular as a shade for cocoa in New Guinea. It has been tried elsewhere but it is not widely used; in Sabah it proved difficult to establish as it requires a well drained soil and will not compete with weeds in its early stages; it also proved difficult to control once established as the common Hawaii variety seeds freely and seedlings develop rapidly where shade is light. *Leucaena* can be thinned with 2.4.5-T.

There are strains of *L. leucocephala*, e.g. Guatemala, which seed less profusely but these have not been widely planted. In Indonesia crosses between *L. leucocephala* and *L. pulverulenta* have given rise to some sterile clones of vigorous growth. There may be a wider use for these in future.

Gliricidia sepium This legume from Central America, where it is called 'madre de cacao', has probably been used as a shade for cocoa longer than any other tree. It is widely distributed and available in most cocoa growing countries.

Gliricidia is planted by using long stakes which are normally easy to establish. It grows quickly up to 9 m (30 ft), and has a fairly light foliage. It loses its leaves during the dry season and flowers, but this can be avoided by lopping shortly before the start of the dry season; the new growth will then hold its leaves. *Gliricidia* is not widely used as a permanent shade but it is sometimes used as shade for the first few years and for shading seed gardens. Growth of *Gliricidia* is easily controlled and it can be killed by 2.4.D. herbicides.

Erythrina species In Trinidad and some other countries in the Caribbean the two species *Erythrina poeppigiana* and *E. glauca* were the traditional shade trees for cocoa, the former being used on higher levels, the latter in lower, wetter places. Both species have been attacked by diseases in Trinidad, the former being badly affected by a witches' broom disease. Trials of alternative shade trees have shown how difficult it is to find suitable shade trees; forty-two species were planted, but only *Inga laurina*, *E. indica* and *E. velutina* seemed worthy of further investigation (Chalmers, 1968).

E. lithosperma is the thornless dadap (*E. indica* is thorny) and was the normal shade for cocoa in Sri Lanka, Indonesia and Samoa. Planted from large stakes, it rapidly provides shade which can be controlled easily. In Sri Lanka and Indonesia dadap was planted at the same spacing as the cocoa but in Samoa cocoa is unshaded and dadap is used as a ground shade, being planted close and regularly slashed to a height of only 75 cm (30 in). It tends to lose its foliage in the dry season but, like *Gliricidia*, this can be prevented by lopping prior to the dry season. Its foliage is liable to attack by insects and where this occurs it is better to plant it in conjunction with another shade tree.

Albizzia species   *Albizzias* are not widely used for cocoa shade but as a genus they have a spreading habit and light feathery foliage providing suitable shade. *A. moluccana* or *falcata* grows very rapidly but is brittle and liable to wind damage. It should be planted at a spacing of at least 18 m (60 ft) and only in places where winds are light. *A. chinensis* grows more slowly and is less brittle but is not so easy to establish. It has found favour in Malaysia. These species are usually grown from seed; hot water or acid treatment of the seeds help to obtain good germination.

Parkia javanica   This tall tree is easy to establish but provides little shade in its early years. In Sabah a trial was laid down comparing growth and yield of cocoa under various shade trees; the yield under *Parkia javanica* was the highest, though this may have been due to a light canopy in the early years (Wyrley-Birch, 1970).

Many other species have been used or tried as shade for cocoa but it is impossible to describe them in detail as most are only of local interest. The brief notes on the main species being used indicate that there are very few trees which can be firmly recommended as shade; it is also true that trees which are suitable in one country may prove a failure elsewhere, *Leucaena leucocephala* being an example. General recommendations, therefore, cannot be made and the choice of shade tree will have to be influenced by local factors, such as the habit of growth, the ease of establishment and subsequent control of growth. It will also be necessary to consider any competition with cocoa, pest attack on the shade tree and whether it is host to any pest or disease of cocoa.

Temporary shade

Other plants are used as temporary shade during the establishment period.

Bananas and plantain   These are commonly used as shade for young cocoa, not because they provide particularly good shade but because they are easily grown and provide food or cash. They are planted at the same spacing as cocoa and will provide shade after six to nine months. There is opposition to the use of bananas and plantains because of their demand for nutrients and moisture. In Nigeria these two crops are not recommended for cocoa shade owing to competition for moisture during the dry season. Analysis of the soil under various shade plants showed consistently lower contents of exchangeable K in the banana and plantain plots (Egbe, 1969). In Trinidad, on the other hand, bananas are recognised as a normal shade for young cocoa. A further disadvantage to the use of bananas and plantains is that, being cash crops, growers are loth to cut them out as the cocoa grows. Development of the cocoa tree is therefore liable to be prolonged and the final result may be unsatisfactory.

Apart from bananas and plantains there are other food crops such as tannias and eddoes, or cocoyams, which are used as ground shade particularly in West Africa. Other crops used less frequently include pigeon pea (*Cajanus cajan*), papaya (*Carica papaya*), cassava (*Manihot esculenta*) and castor oil (*Ricinus communis*).

Tree cassava or Ceara rubber (Manihot glazovii)   This plant is easily grown from stem cuttings which should be planted at least 0·3 m (1 ft) deep in order to ensure good anchorage. It grows to a height of 4 to 6 m (15 to 20 ft) and forms a fairly heavy canopy. In West Africa it is considered to be a useful plant for filling gaps in forest shade and is used as a temporary shade for this purpose. In the Far East it has been used for the same purpose but has proved attractive to pigs which dig out the tubers; it can also be difficult to eradicate.

Gliricidia and dadap (*E. lithosperma*), already described as permanent shade, are used as temporary shade in some countries, dadap being used in this way in Samoa.

Cover crops

The problems involved in clear-felling and consequent exposure of the soil might be overcome by the use of a quick-growing cover crop but they have found little favour in cocoa growing except in a few countries, New Guinea for instance. This is partly because cover crops will have a short life under cocoa, being suppressed as the canopy closes; it is also because cover crops have their disadvantages: they may be difficult or expensive to establish, they may be ineffective or, alternatively, too effective, covering the cocoa as well as the soil, or they may suffer from or encourage pests and diseases.

Cover crops have not been used in West Africa because cocoa is generally planted under thinned jungle but their effect on soil and the growth of cocoa has been studied in Ghana and Nigeria (Cunningham and Smith, 1961; Jordan and Opoku, 1966; Longworth, 1963). The trials in West Africa compared natural regeneration with bare soil, mulches and various crops and showed that natural regeneration, which is frequently used to provide initial shade, is the least satisfactory method. Bare soil, by removing all competing vegetation, allows the cocoa seedlings to use all the moisture and nutrients available, but is liable to erosion by heavy rain, and the maintenance of such conditions is expensive. Mulches usually give the best results but the expense of growing, cutting and spreading a mulch is generally considered pro-hibitive. Cover crops are a more practical alternative giving better results in terms of growth and yield than natural regeneration though less effective initially than a mulch.

Of the cover crops tested in West Africa, *Crotalaria striata* and *Pueraria javanica* were the most satisfactory for ease of establishment and effectiveness of cover. Cover crops which establish quickly tend to die out quickly; in the case of the two species mentioned, *C. striata* establishes more quickly than *P. javanica*.

In Malaya (Mainstone, 1971), *C. striata* and *C. anagyroides* have been found to spread too much and need slashing. Another erect cover, *Sesbania punctata*, is useful, establishing easily and growing quickly to a height of two metres to provide lateral shade for about 18 months. *Tephrosia candida* and *T. vogelii* have also been tried, but establish more slowly than *S. punctata*.

In Papua–New Guinea (Henderson, 1954) *Crotalaria anagyroides* is the most commonly used cover but it is susceptible to pink disease (*Corticium salmonicolor*) and is easily blown down. It lasts for about 18 months. To provide a longer period of cover and some support for *C. anagyroides* it is sometimes mixed with *Tephrosia candida*.

Sequence of operations

The sequence of operations when planting cocoa after clear-felling is, therefore, as follows:

> Underbrush
> Layout roads and dig drains where necessary
> Fell jungle
> Cut up the large tree trunks
> Stack and burn
> Line and plant shade; this operation should follow quickly after
>   stacking and burning to reduce weed competition
> Plant cocoa when shade conditions are right

Where there is a marked dry season the felling should be done near the beginning of the dry season and the burning at the end so that the shade can be planted as soon as the rains start. This may mean as much as a year's delay before the cocoa can be planted, as cocoa will establish more satisfactorily when planted in the early part of the wet season. Where wet and dry seasons are less pronounced, as in the Far East, cocoa can be planted in most months of the year. The programme of land preparation and planting can, therefore, continue through several months and there will be a shorter interval between clearing and planting cocoa.

There are many variations, major and minor, to this sequence of operations. The most important variation is to remove the jungle trees by poisoning, having first established permanent shade and cocoa. This can follow clearing of the undergrowth with or without a light burn. This method has been employed in Papua–New Guinea (Newton, 1966; Blow, 1968) and has the advantage of saving both time and money but there may be difficulties with falling trees.

Planting pattern

In establishing cocoa after clear-felling it is necessary to devise an arrangement of temporary and permanent shade which will give the changing degree of shade required for optimum growth of the cocoa. Innumerable patterns have been evolved, the simplest being a hedge of *Leuceana leucocephala* interrupted for the cocoa hole which is surrounded by *Crotalaria anagyroides*—this is used in Papua–New Guinea. Most planting patterns are more complicated as they include a mixture of species for both permanent and temporary shade. One or two permanent shade trees may be planted at 12×12 m or 18×18 m (40×40 ft or 60×60 ft); temporary shade at the same spacing as for the cocoa seedlings, and cover crops may be planted around the cocoa planting space or in the inter-row. An example of such a planting plan is shown in Fig. 7.1. In designing a planting pattern access to the mature cocoa trees for spraying and harvesting must be considered; for convenience the cocoa and the permanent shade trees should be in the same row.

Jungle shade

Thinning of the jungle and planting of cocoa beneath the remaining jungle trees is the method commonly used in West Africa, but the method varies according to the soil, climate and distance from markets. In places where there is no shortage of soil moisture during the year, the jungle is thinned leaving five or more dominant trees and 35 to 45 smaller trees per hectare (Hammond, 1962). In Nigeria and the drier areas of Ghana the forest may be completely felled as the farmers have learnt that the land will not support forest shade and cocoa during the dry season and that the cocoa suffers under such conditions. It is normal to plant food crops—plantains and cocoyams—as nurse crops for the

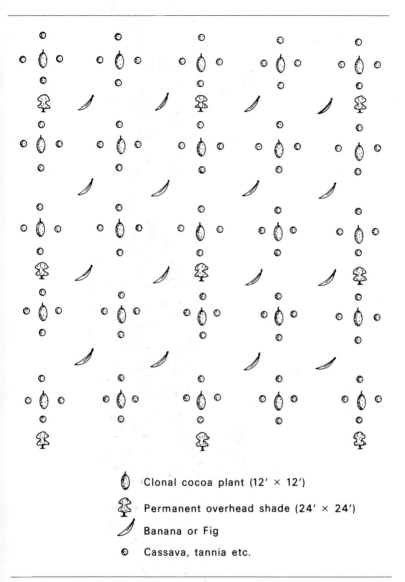

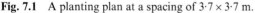

Fig. 7.1 A planting plan at a spacing of 3·7 × 3·7 m.

cocoa seedlings. Where the farm is remote and there is no market for
food crops, the cocoa may be planted without any nurse crop.

Farmers will select the trees to be left for shade partly according to
the shade they provide but also to some extent according to the work
involved in felling. Some particularly large trees may be left simply

because they are too laborious to fell. On the other hand, some tree species are removed because they are not compatible with cocoa. This aspect was studied in the Congo and lists of compatible and incompatible trees have been published (Poncin, 1958).

The advantages of this method lie in its speed and economy. It is possible to save a year in establishing cocoa by underplanting jungle, the usual programme being to underbrush early in the dry season, line and peg, fell any surplus jungle trees to achieve the right degree of shade and to clear the planting lines. The land should then be ready for planting cocoa as soon as the ground is moist enough. The economies that result from this method are obvious. There are, however, disadvantages. The jungle shade is never uniform and is more difficult to control and adjust than planted shade. The jungle trees do not, of course, fall into planting lines and this can be a hindrance where it is desirable to move between the rows with a tractor or some forms of spraying machine. This method is used almost exclusively in West Africa where the jungle trees provide suitable shade and where its advantages are of paramount importance to the small farmers; it is also used in Brazil. In Malaysia this method is not always successful because the jungle trees fail to provide suitable shade, their crowns being dense and small.

In place of the normal felling of jungle trees by hand, the jungle can be cleared or thinned more cheaply by poisoning the trees with 2.4.5-T, a technique that has been tried experimentally in Ghana (Liefstingh, 1966). After initial trials an area of 7 hectares was poisoned and at the same time planted with cocoa seed at stake at a spacing of 3×0.6 m (10×2 ft), the dense spacing being adopted in order to minimise damage at a later stage from falling branches and tree trunks. The total costs for poisoning and planting was very much less than the cost of usual methods. The poisoning involved the use of 2·5 per cent solution of 2.4.5-T in diesel oil painted on to the trunk of the tree over a band about 0·3 m (1 ft) wide. Poisoning 200 trees required 8 gallons (36 litres) of this mixture. Most of the trees were dead within a year and the cocoa seedlings developed well with little loss from falling trees. Poisoned trees tend to fall piecemeal causing less damage than felling. In this trial this technique was used to clear land completely, but it could also be used to thin jungle.

Interplanting

Establishing cocoa is easiest where it is possible to underplant another crop but the only crop providing suitable shade for mature cocoa is the coconut palm. Interplanting coconuts with cocoa is a long-established practice in Papua–New Guinea which was extended rapidly after the war; more recently, the practice was taken up on the West Coast of Malaya and in Sarawak.

Interplanting coconuts should only be adopted where soil conditions are suitable for cocoa. In many countries coconuts are grown on sandy soils which are quite unsuitable for cocoa. In Papua–New Guinea, however, coconuts are grown on pumice soils and in Malaya on coastal clay soils; these soils are suitable for cocoa provided there is well distributed rainfall on the free-draining pumice soils and good drainage on the coastal clays.

The practice of interplanting mature coconuts with cocoa is attractive because the cost of establishing cocoa is low, the income per hectare is increased and the cost of maintaining the coconut area is reduced. Furthermore, the two crops are compatible; there is no indication that the presence of one crop reduces the growth of the other and they do not share common pests and diseases.

In Papua–New Guinea the coconuts are usually spaced at 9 m (30 ft) and the cocoa was planted at 4·5 m (15 ft) intervals between and within the rows of palms; this gives 360 cocoa trees per hectare. The young seedlings are given temporary additional shade of palm fronds or *Crotalaria anagyroides* where the palms are at a wider spacing. It is found desirable to dig a large planting hole in order to break up the mat of coconut roots.

In Malaya coconut palms are planted 8 or 9 m (26 or 30 ft) apart and two rows of cocoa seedlings are planted between the rows of palms. The spacing between the cocoa trees varies but a planting density of 1,040 trees per hectare (435 trees per acre) is the usual aim. No cocoa trees are planted in the rows of palms, because this would interfere with coconut harvesting which, in Malaya, is done by means of long harvesting poles.

Before planting the interrow with cocoa the soil is sometimes roto-vated in order to reduce weed growth and break up the coconut roots. This is particularly useful when seed at stake is planted. Extra shade is often required at planting and this is easily provided by using coconut fronds, while coconut husks are placed around the seedlings to form a mulch and reduce weed growth close to the plant (Chalmers, 1968).

There have been attempts to interplant rubber and oil palms with cocoa but in neither case is it successful. At normal spacings both crops provide too heavy shade for cocoa which may develop vegetatively but will not bear an economic crop. There are, however, special circum-stances in which rubber can be interplanted with cocoa. In the Matale district of Sri Lanka old rubber was interplanted, but this district is basically unsuitable for rubber as conditions favour the leaf disease caused by *Oidium heveae*. The old rubber was thinned to only 160 trees per hectare and cocoa planted at 4·5 × 3 m (15 × 10 ft). In these special circumstances cocoa has been grown successfully.

Another possibility is to plant cocoa as a sort of catch crop during rubber replanting. This would involve planting cocoa in old rubber due for replanting. The rubber would be felled or poisoned shortly after the cocoa is planted and the latter may then require some additional shade. The land would be replanted with rubber at 12×2 m (40×6 ft) but several crops of cocoa could be harvested before the rubber overshades the cocoa. It has been suggested that this method would help to promote rubber replanting on smallholdings in Malaya by providing some income while the new rubber is developing (Blencowe, 1968).

Other objections, apart from shade, have been raised to the interplanting of rubber with cocoa. There is the possibility of root disease causing losses from both crops and the aggravation of panel diseases and pod diseases owing to a more humid microclimate. The interplanting of rubber with cocoa has never been extensive enough to test these points.

In Western Nigeria cocoa farms have been interplanted irregularly with kola (*Cola nitida*) and in Grenada nutmegs are planted among cocoa trees. In both cases the cocoa becomes heavily overshaded by the other crop.

Windbreaks

In countries where winds blow steadily during certain months, for instance the West Indies where the south-east trades wind blows during the first part of the year, windbreaks are usually planted.

In Grenada the cocoa fields, which are only a few acres in extent, are unshaded but surrounded by mango or galba (*Calophyllum antillanum*). The mangoes are planted very close, $0 \cdot 3$ to $0 \cdot 6$ m (1 to 2 ft) between each plant, and sometimes in a double row. Many plants die but the resulting hedge is fairly dense. Similar hedges are planted in Jamaica. In Trinidad it is commoner to use *Dracaena* or *Hibiscus* which form thick hedges, 3 to $4 \cdot 5$ m (10 to 15 ft) high. In Fiji cloves (*Eugenia aromatica*), Malacca apple (*Eugenia malaccensis*) and mahogany (*Swietenia macrophylla*) are used as well as mango. In Samoa teak (*Tectona grandis*) has been planted as a windbreak, and in Zanzibar cinnamon is used.

There is thus a considerable variety of trees and plants used as windbreaks. The choice of plant will be influenced by its local growth habit in relation to the wind and the time of year when protection is required. A windbreak must be aligned suitably in relation to wind direction and local topography, and the distance between the windbreaks will have to be related to the height of the windbreak and the strength of wind; a windbreak affects the wind velocity at distances up to six times its height.

Spacing

The optimum spacing between cocoa trees is the distance which will give the greatest economic return of cocoa per unit area. It will be affected by several factors; the vigour of the trees, type of planting material, shade conditions, soil and climate.

Each country has adopted a certain spacing which has become traditional. In Sri Lanka, New Guinea and Samoa the spacing is usually 5×5 m; in Trinidad, the Dominican Republic and Central America the common spacing is 4×4 m, while in South America a rather closer spacing of 3 to 4 m is employed. West Africa is the only area where really close spacing is used; two or three seeds are planted close together and the groups of seeds are about 1·3 m apart in a haphazard manner. In this case many trees die before they reach maturity and the final stand is commonly of the order of 1,500 trees per hectare (600 trees per acre) equivalent to a spacing of $2·5 \times 2·5$ m. As the farmers do not select the seedlings as they develop, many twin trees can be found in West African farms.

Most of the experimental evidence points to close spacing giving the highest yields. There is no doubt that a close spacing will give a higher yield in the early years but once a canopy forms and the soil becomes fully exploited the difference between close and wide spacing narrows.

This is shown in the results of various spacing trials, for instance, a shade, spacing and fertiliser trial planted at River Estate, Trinidad in 1949. This trial gave the following yields:

Table 7.1 *Shade and spacing trial, River Estate (yields in kg per hectare—unshaded plots)*

Years	$2·7 \times 2·7$ m $(8 \times 8 ft)$	4×4 m $(12 \times 12 ft)$
1952–53	384	205
1953–54	473	390
1954–55	1,349	1,048
1955–56	1,405	1,102
1956–57	1,520	1,318
1957–58	1,180	1,292

SOURCES: Havord *et al.*, 1954, 1955; Maliphant, 1959.
Original data converted to metric.

While the yields at the end of this period were similar for the two spacings, the close spaced plots had given an aggregate of 6,311 kg against 5,355 kg for the wide spaced plots. On the shaded plots the yields were similar but the differences between close and wide spacings were smaller.

In West Africa spacing trials have been planted at Tafo in Ghana and at Gambari in Nigeria. In both instances, a close spacing of $2·3 \times 2·3$ m $(7·5 \times 7·5$ ft) gave high yields; closer spacings have higher yields in some years, but not significantly more.

At Gambari the trial compared seven spacings from $1·5 \times 1·5$ m to

4·5 × 4·5 m (5 × 5 ft to 15 × 15 ft). Kowal (1959) discussed this trial in detail and concluded that a density of less than 1,440 trees per hectare (2·6 × 2·6 m) 'is disadvantageous in relation to health and management and uneconomic in relation to yield'.

At Tafo two spacing trials have been planted with Amelonado and Amazon cocoa respectively. In the Amelonado spacing trial planted in 1947 eight spacings from 1·2 up to 4·6 m (4 × 4 ft up to 15 × 15 ft) were compared. A spacing of 2·3 × 2·3 m (7·5 × 7·5 ft) gave the highest overall yield, the aggregate yields for eight seasons to 1961–62 being (Wood, 1964):

metres	ft	kg per hectare
4·6 × 4·6	15 × 15	7,590
3·7 × 3·7	12 × 12	10,396
3·0 × 3·0	10 × 10	12,209
2·3 × 2·3	7·5 × 7·5	13,324
2·3 × 1·8	7·5 × 6	11,894
1·8 × 1·8	6 × 6	11,806
1·5 × 1·5	5 × 5	12,533
1·2 × 1·2	4 × 4	12,105

Apart from the yield at 2·3 × 2·3 m, all spacings from 3 × 3 m down to 1·2 × 1·2 m have given the same yield.

In the Amazon spacing trial, the trees were planted at 2·4 × 2·4 m, 3 × 3 m and 3·7 × 3·7 m. The results have not followed quite the same pattern, the closer spacing gave slightly greater yields in the early years but from eight years on the wider spacings have yielded more heavily. After twelve years the total yields were (Wood, 1964):

metres	ft	kg per hectare
2·4 × 2·4	8 × 8	8,804
3·0 × 3·0	10 × 10	9,710
3·7 × 3·7	12 × 12	9,663

In Papua–New Guinea the usual practice has been to plant cocoa at a spacing of 4·6 × 4·6 m (15 × 15 ft) and this has been the recommended distance. Charles (1961) has reported that in trials at Keravat comparing a spacing of 3·6 m triangular with 4·6 m triangular, the closer spacing has yielded 17 per cent more cocoa over a period of eight years and suffered fewer casualties. Closer spacings than 3·6 m have not been tested.

The results of all these trials point to a spacing between 3 × 3 m and 2·3 × 2·3 m as giving the highest yield. At this spacing the canopy forms fairly quickly, thereby reducing weed growth and weeding costs; losses from certain pests appear to be appreciably lower. On estates and plantations 3 × 3 m is a convenient spacing which allows easy access to the rows of trees and the possibility of using a tractor between the rows where the land is reasonably flat. If a closer spacing is desired then a spacing of 3×2.4 m might give slightly higher yields while still giving easy access in one direction.

Under humid conditions it may be desirable to reduce the canopy and this could be done by increasing the distance between the rows and planting closer within the row. This might reduce humidity and hence the losses to pod diseases; a suitable spacing might be 3.7×2.4 m (12×8 ft) but this has not been tested experimentally.

Apart from the difficulty of access where very close spacings are used, they involve the planting of large numbers of plants. When the cost of a seedling together with the cost of staking, holing and planting is 5p a plant, the cost per acre will be £22 at 3×3 m, £41 at 2.3×2.3 m and £61 at 1.8×1.8 m. The higher capital costs must be recouped by higher yields or lower maintenance costs. At spacings closer than 3×3 m these gains appear to be marginal.

There have been occasions when cocoa has been planted close and thinned later. The object of this operation is to achieve an early yield or to reduce maintenance in the early years. The drawbacks to this method are economic and psychological. As explained earlier, the additional capital costs in planting close are considerable and there is no evidence to show whether they are equalled by extra yield or lower costs. The psychological difficulty lies in the fact that there is some reluctance to thin out trees once they start bearing, and, where a regular pattern is the aim, this is bound to involve removing some vigorous trees and leaving some less vigorous ones.

Holing and planting

In some countries it is customary to dig a planting hole for cocoa as for other permanent crops. This has been a common practice in the West Indies and in the Far East. In Trinidad the usual practice is to dig a hole $40 \times 40 \times 24$ cm deep, mix the soil with pen manure and return it to the hole at planting time (Havord, 1953). Where the soil is heavy the holes are dug several months or weeks before planting in order to weather the exposed soil. This practice has been recommended in several countries including New Guinea where the planting holes are 45 cm square and 60 cm deep; in Fiji they are 60 cm square and 30 cm deep (Hardy, 1960, p. 28; Harwood, 1959; Henderson, 1954). The extreme case is the practice that was carried out in São Tomé where planting holes 45 to 60 cm in diameter and 75 cm deep were dug in good loose soil and up to 200 cm deep in heavy or stoney soil (van Hall, 1932). At the other end of the scale the West African farmer never digs a planting hole for seed, while, for seedlings, the hole is no bigger than is required to hold the plant and its ball of earth; holing, under these circumstances, is not a separate task—the hole being made at the same time as planting.

A trial conducted in Trinidad compared planting holes 30, 60 and 100 cm square by 15, 30 and 60 cm deep respectively, the soil being mixed with proportional quantities of pen manure. No significant differences were found in girth and yield of the plants but unfortunately the plants tended to become waterlogged because the planting hole formed a basin as the manure decomposed (Murray and Maliphant, 1964).

There is clearly a considerable amount of labour involved in preparing planting holes—12 to 13 man-days per acre has been quoted for lining and holing in Trinidad and Grenada—and this must be justified in terms of better growth and subsequent yield. The current practice seems to be closer to that of the West African farmer, small holes being dug at the same time as planting, just large enough to contain the seedling and the soil in its basket or pot. There seems to be no real evidence that any other form of treatment is necessary on soils of reasonably open texture. Where the soils are particularly stoney or heavy some benefit will be gained by digging a planting hole and improving the soil by removing the stones or weathering the clay.

The findings of the experiment quoted earlier do not give a guide as to the value of pen manure in the planting holes. It seems likely that this practice is declining owing to expense or shortage of pen manure, and, in the absence of experimental evidence, the continuance of this practice, however beneficial it might appear to be, cannot be firmly recommended.

Time of planting

The soil must contain adequate moisture for young cocoa plants at the time of planting and during the following months whilst the tap root is developing. In countries with a pronounced dry season the plants must be firmly established before the dry season starts; this means that the tap root must have grown to a depth at which moisture can be obtained throughout the dry season. The inference is that cocoa should be planted at the beginning of the wet season, rather than at the end, and this is the general rule in West Africa. There are countries which are fortunate enough to have suitable conditions during most of the year and this allows a more or less continuous planting programme which has obvious advantages.

Method of planting

Where seed is sown at stake it is only necessary to make a small hole sufficient for one seed, and two or three such holes close together for each picket. When planting seedlings from a nursery it is necessary to ensure that shade conditions in the nursery are approximately the same as in the field prior to planting. If there is a great difference then shade should be reduced in the nursery, or supplementary shade—palm fronds, for instance—can be used in the field.

The plants should be well watered before being carried to the field and precautions should be taken to ensure that the soil around the plants is not shaken severely during transit to the planting site. This will help to prevent any damage to the tap root at the time of planting. It is easy for the tap root to be bent at planting and this will lead to the plant being physically weak and unable to withstand winds or drought. This is probably the most important point to be considered when planting seedlings.

References

Blencowe, J. W. (1968) 'Cocoa growing under rubber: the prospects', *Cocoa and Coconuts in Malaya*, Symposium Incorp. Soc. Planters, Kuala Lumpur, 1967, pp. 57–60.

Blow, R. (1968) 'Establishment of cocoa under jungle and conversion to planted shade', *Cocoa Growers' Bull.*, **11**, 10–12.

Chalmers, A. (1968) 'Establishing cocoa under coconuts; the early stages', *Cocoa and Coconuts in Malaya*, Symposium Incorp. Soc. Planters, Kuala Lumpur, 1967, pp. 12–19.

Chalmers, W. S. (1968) 'Shade trees for cacao', *Ann. Rep. Cacao Res. 1967*, Trinidad, pp. 47–50.

Charles, A. E. (1961) 'Spacing and shade trials with cacao', *Papua and New Guinea Agric. J.*, **14**, 1–15.

Cunningham, R. K. and Smith, R. W. (1961) 'Comparison of seed covers during cocoa establishment on clear-felled land', *Trop. Agric.*, **38**, 13–22.

Dun, D. S. (1967) 'Cacao flush defoliating caterpillars in Papua and New Guinea', *Papua and New Guinea Agric. J.*, **19**, 67–71.

Egbe, M. E. (1969) 'The effects of some temporary shade plants of cacao on certain nutrient contents of the soil', *Proc. 2nd. Inter. Cacao Res. Conf. Bahia 1967*, pp. 333–4.

Freeman, G. H. (1964) 'Present nursery and establishment methods for cocoa in Western Nigeria', *Ann. Rep. W. Afr. Cocoa Res. Inst. (Nigeria) 1962–63*, pp. 13–24.

Hall, F. J. J. van (1932) *Cacao*, Macmillan, p. 459.

Hammond, P. S. (1962) 'Cocoa agronomy', in J. B. Wills, ed., *Agriculture and Land Use in Ghana*, Oxford University Press, pp. 252–6.

Hardy, F., ed. (1960) *Cacao Manual*, Turrialba, Costa Rica, Inter-American Inst. Agric. Sciences, p. 28.

Harwood, L. W. (1959) 'Cocoa planting', *Agric. J. Fiji*, **29**, 65–75.

Havord, G. (1953) 'Manurial and cultural experiments on cacao', *Rep. on Cacao Res. 1945–51*, Trinidad, pp. 104–8.

Havord, G., Maliphant, G. K. and Cope, F. W. (1954) 'Manurial and cultural experiments on cacao Part III', *Rep. on Cacao Res. 1953*, Trinidad, pp. 80–7.

Havord, G., Maliphant, G. K. and Cope, F. W. (1955) 'Manurial and cultural experiments on cacao Part V', *Rep. on Cacao Res. 1954*, Trinidad, pp. 65–8.

Henderson, F. C. (1954) 'Cacao as a crop for the owner-manager in Papua and New Guinea', *Papua and New Guinea Agric. J.*, **9**, 45–74.

Jordan, D. and Opoku, A. A. (1966) 'The effect of selected soil covers on the establishment of cocoa', *Trop. Agric.*, **43**, 155–66.

Kowal, J. M. L. (1959) 'The effect of spacing on the environment and performance of cacao under Nigerian conditions. I. Agronomy', *Emp. J. Exp. Agric.*, **27**, 105, 27–34.

Liefstingh, G. (1966) 'Is chemical clearing a possibility?', *Cocoa Growers' Bull.*, **6**, 12–16.

Longworth, J. F. (1963) 'Improved methods of establishing cocoa on clear-felled land in Nigeria', *J. Hort. Sci.*, **37**, 222–31.

Mainstone, B. J. (1971) 'A background to Dunlop work with covers and shade for cocoa', *Cocoa and Coconuts in Malaysia, Proc. Conf. Incorp. Soc. Planters, Kuala Lumpur 1971*, pp. 102–11.

Maliphant, G. K. (1959) 'Manurial and cultural experiments on cacao Part VII', *Rep. on Cacao Res. 1957–1958*, Trinidad, pp. 83–6.

Murray, D. B. (1953) 'A shade and fertilizer experiment with cacao. Progress report—continued', *Rep. on Cacao Res. 1952*, Trinidad, pp. 11–21.

Murray, D. B. and Maliphant, G. K. (1964) 'Size of planting hole', *Ann. Rep. Cacao Res. 1963*, Trinidad, pp. 49–50.

Newton, K. (1966) 'Methods of establishment and shade management', *Tech. Mtg on Cocoa Prodn, Honiara*, S. Pacific Comm. Paper No. 19.

Poncin, L. (1958) 'The use of shade at Lukolela Plantations', *Rep. Cocoa Conf. London 1957*, pp. 281–8.

Wood, G. A. R. (1964) 'Spacing', *Cocoa Growers' Bull.*, **2**, 16–18, and note in *Cocoa Growers' Bull.*, **3**, 28.

Wyrley-Birch, E. A. (1970) 'Shade for cocoa', *Cocoa Seminar, Tawau 1970*, Sabah Planters Assoc., pp. 51–9.

Chapter 8

Maintenance and Rehabilitation

Maintenance and rehabilitation are quite distinct aspects of cocoa cultivation, differing in the initial condition of the cocoa. Maintenance is concerned with the growth and development of young cocoa and keeping mature farms in the right condition for optimum yield. Rehabilitation is concerned with old, neglected, diseased or low-yielding farms which can be restored to healthy farms giving a good yield.

Maintenance

The object of the operations involved in maintaining a cocoa farm is to provide optimum conditions for growth and yield; to this end shade must be adjusted, weeds, pest and diseases kept under control, the cocoa trees pruned and fertilisers applied. The adjustment of shade is mentioned in Chapter 9; this will be a regular task as the shade should be reduced while at the same time shade trees grow. The thinning of shade trees is best done during the dry season and by poisoning as described in Chapter 7. The control of pests and diseases is dealt with in Chapters 10 and 11 and the approach to the use of fertilisers in Chapter 9.

Weeding

The purpose of weeding is to reduce competition, to prevent weeds from climbing up the trunks of cocoa trees, to allow access to the trees, and hence to make the tasks of spraying and harvesting easier.

There is not much experimental evidence to show the effect of weed competition on the growth of seedlings. Walmsley (1964) reported a trial in which some plots were kept bare of weeds by use of herbicides, and on other plots weeds were slashed once a year. There was a significant difference in the number of seedlings that needed replacement and in mean trunk diameter, the advantages being with the weed-free plots. Similar results were obtained from some trials conducted in Ghana on sandy and sandy loam soils which are more subject to drought than the better cocoa soils (Brown and Boateng, 1972a). In this case spraying with paraquat was compared with different methods of brushing with a cutlass.

While there is no other direct evidence on the competitive effect of weeds, there is the important indirect evidence of the correlation between tree size and yield. Tree size reflects the growing conditions during the early years in the field. Jones and Maliphant (1958) studied the results of long-term fertiliser trials and one of their conclusions was that 'if weed competition and soil moisture deficits are not controlled during the early stages of establishment . . . then extreme variability in yield can be expected later'.

The control of weeds is the most demanding of the various operations, as frequent weeding is usually necessary. The rate of weed growth depends on local conditions of shade, climate and season, but monthly weeding rounds are needed during the wet season in most countries. This is certainly true of young farms but the amount of weed growth should diminish as the canopy develops.

The cutlass remains the usual tool for weeding, but must be used with care to avoid damage to young seedlings or to the root systems of cocoa trees. Seedlings can be protected either by hand weeding in the area close to the seedling or by placing a mulch of cut weed growth around the stem, which will also reduce weed growth close to the seedling. Root systems are preserved by slashing the weeds without scraping the top soil.

In young cocoa weeding consists of slashing interrow growth, some of which may be used as a mulch, together with the removal of creepers. As the seedlings develop and a canopy forms the intervals between the rounds of weeding can be increased, alternatively the regular frequent rounds will not require as much labour. In more mature plantings weeding is often combined with the removal of basal chupons. In mature cocoa weed growth is suppressed by the heavy shade beneath the canopy. Regular weeding is still necessary but the task becomes a light one involving little labour for each cycle.

Herbicides

Herbicides are a useful substitute for the cutlass where weed growth is severe, provided they can be applied safely. Herbicides are classified by their mode of action, and are divided into contact and translocated herbicides, the latter being subdivided into those which are absorbed through the leaves and those taken up by the roots.

Paraquat is a commonly used contact herbicide. It is highly effective in killing off top growth, but is not absorbed by woody tissue and has no effect on a deep-rooted plant, so that woody weeds are less affected, and can become a nuisance. To overcome this, paraquat is now more commonly used in mixtures with other herbicides. Paraquat must not be sprayed on to cocoa leaves or green wood, but will not affect cocoa plants if it falls on mature wood.

The phenoxyacetic acids (2.4-D and 2.4.5-T) and the halogenated aliphatic acids (dalapon, TCA) are translocated herbicides absorbed through the foliage and are effective against annual and perennial broadleaf weeds and grasses. The ureas (e.g. diuron and linuron) and the triazines (e.g. atrazine and simazine) are absorbed through the root system. All the translocated herbicides must be kept off the leaves and growing points of the crop being grown, and 2.4.5-T must be kept off the woody parts as well.

The effect of herbicides on the growth of cocoa has been studied by Kasasian and Donelan (1965) in Trinidad and by Brown and Boateng (1972b) working in Ghana. The Trinidad trials showed that it is safe to use 3·3 kg per hectare of simazine and linuron for long term control and for short term weed control 1·1 kg per hectare paraquat, 2·2 kg per hectare 2.4-D, 2.4.5-T or MCPA and 5·6 kg per hectare dalapon. Results with diuron have been conflicting, damage having been reported in some cases. The trials in Ghana produced similar results and tested some additional herbicides; MSMA, an organic arsenical compound of low toxicity, appeared to produce leaf symptoms similar to those of zinc deficiency, but this has not been reported from Malaya where this herbicide has been used quite extensively.

Various recommendations have been made for weed control in cocoa. Table 8.1 incorporates suggestions made for the Caribbean area.

In Malaya the use of weedkillers is possibly more advanced than anywhere else in the tropics, and effective herbicide mixtures have been evolved to cope with the difficult weed conditions that occur there. The following mixture has been recommended for use where cocoa is planted under coconuts on coastal clay soils—conditions of light shade and continuously wet soil (Evans, 1970):

	Imperial per acre	Metric per hectare
MSMA	5 pints	7 litres
2.4-D	1 pint	1·4 litres
Sodium chlorate	5 lb	5·6 kg
Water	40 gallons	450 litres

After making up this mixture a wetting agent is added. The above volumes are applied as an initial treatment prior to planting. The volume is reduced for subsequent treatment at monthly intervals to 75 per cent and 50 per cent of the initial figure. This is achieved by use of different nozzles. After the third treatment the sodium chlorate is replaced by 2·2 kg (2 lb) of dalapon to control grasses.

Table 8.1 *Suggested herbicide treatments*

Weeds controlled	Chemical	Rate	Usual period of control (weeks)	Remarks
	Paraquat	1–2 pts	3–6	
Annual and perennial broadleaf weeds and grasses	Paraquat + Atrazine or Diuron	1 pt 3–4 lb 1–2 lb	6–12	Keep off leaves and growing points
Perennial grasses mixed with broadleaf weeds	Dalapon + 2.4-D + Atrazine or Diuron	3 lb 1 pt 3 lb 2 lb	10–20	Keep off leaves and growing points.
	Dalapon + 2.4-D	5 lb 1–2 pt	8–16	
Annual and perennial herbaceous broadleaf weeds	2.4-D	2–3 pt	6–12	Take great care to avoid spray drift
Annual and perennial woody broadleaf weeds	2.4.5 -T	2–4 pt	6–12	
Annual and perennial grasses	Dalapon	5–10 lb	8–16	

SOURCE: Kasasian (1965).

This recommendation may be of local importance, but it is described because it illustrates certain principles that are common to many situations. In the first place it is only in the years of establishment that herbicides are going to be used in any quantity. Thereafter weed growth should be reduced to a low level by the cocoa canopy and any trouble-some weeds can be dealt with by spot treatment with a herbicide or by hand weeding. Second, the mixture is designed to deal with the local weed population, and the equipment is selected so as to apply the mixture safely and at the correct dosage. Third, a fairly heavy dose is used initially but this can be reduced in quantity in successive treat-ments or the interval between treatments can be increased. Finally, the weed population will almost certainly change and this will require changes in the herbicide mixture.

Herbicides are being developed continuously and changes take place frequently, so that firm recommendations would be out of place. On the other hand, the techniques for herbicide application have developed to an advanced stage and have been described in detail by Shepherd *et al.* (1970).

Pruning

There are several objectives in pruning:

1. To shape the developing tree.
2. To ensure easy access to the trees for spraying and harvesting.
3. To help control pests and diseases.
4. To ensure high yields and optimum return from the tree.

The only pruning required by small seedlings is to remove one stem from trees with double stems. Seedlings grow vertically until they jorquette to form three to five fan branches. The height at which seedlings jorquette varies considerably, the usual range being between one and two metres. The stimulus to form a jorquette is unknown, but light seems to have some bearing on it. In strong light seedlings jorquette at a lower height than in heavy shade, but the height of jorquettes is always very irregular. In Malaya the jorquette height has been adjusted to an even height by pruning off jorquettes which form below a certain level (Leach *et al.*, 1971). This is done monthly and results in several chupons being formed from buds below the jorquette point; the strongest chupon is selected and the others removed. While the young chupons may be weak, the stem becomes straight and the junction strong.

The number of fan branches formed at a jorquette varies; it is usually four or five, sometimes only three. When five branches are formed one is usually weaker than the rest and planters in some countries remove the weaker fan branches.

In mature fields easy access can be ensured by the removal of low drooping branches. The third objective is partly sanitation, and the removal of any branches or other parts of a tree damaged by pest and disease and likely to provide cause for either to spread is only a commonsense precaution. In many countries this is all that will be necessary to meet this objective but further action may be advisable where pod diseases occur.

Pruning to help control black pod is problematical. In Fernando Po the number of fan branches is reduced to three, the trees are restricted to the first jorquette and chupons are removed from the trunks (Swarbrick, 1965). A closed canopy is maintained and the jorquette is shaded so this method helps to control black pod by easing the task of spraying. In other countries the canopy is kept open or the centre of the tree opened up by pruning in order to reduce humidity and hence reduce black pod infection. The effectiveness of this practice in reducing black pod infection is uncertain but severe pruning to 'open up' the trees can merely lead to loss of yield (Newton, 1966).

After the formation of the jorquette, a decision must be made whether to allow a chupon to develop below the jorquette to form a second storey, but there is no clear evidence of the effect of this on yield. The merits of restricting the growth of cocoa trees to the height of the first jorquette have been widely discussed. The advantages are claimed to be easier harvesting and maintenance of the trees by restricting their height, but the value of these advantages, the cost of pruning, and its effect on yield are uncertain because little experimental work has been done on this aspect of cocoa cultivation.

Bonaparte (1966) has reviewed the experiments that have been conducted in Ghana. In one of these trials the costs of harvesting were measured, and in terms of pods harvested per man-day showed that the difference in cost was negligible. Therefore the additional cost of regular pruning would not be recouped from cheaper harvesting, but would have to depend on increased yield. Trials were conducted to compare restriction to the first jorquette with no pruning. In the early years the pruned trees yielded slightly more than the unpruned trees, but, after ten years from planting, the unpruned trees started to yield more heavily than the pruned trees. This applied to both Amazon and Amelonado whose habits of growth are different and whose response to pruning might be expected to differ. In another trial a more flexible pruning regime was tried, in which a second jorquette was allowed to develop when the first was formed below a certain height. In this trial the pruned trees gave higher yields, but the length of the trial was unsatisfactory and the result was not significant.

To sum up, pruning of mature cocoa consists of the regular removal of basal chupons, the removal of low branches and all dead wood; where black pod occurs further action should be taken to make spraying simpler and possibly to lighten the canopy; finally a decision must be made about the second and later jorquettes. Many planters restrict growth to the first jorquette but there is little evidence for or against this practice.

Supplying

It is customary to supply vacancies in a field of young cocoa. In the first year the proportion of dead and missing trees may be 10 per cent but should not exceed 15 per cent. At this stage all vacancies should be supplied by new seedlings, but after the first year it becomes progressively more difficult to look after individual seedlings within a field of more mature trees and to ensure that they are provided with suitable shade. Therefore, the practice of repeated supplying of vacancies after the first year where growth is rapid, and at a later stage where growth is slower, cannot be recommended. It is only where large gaps occur in a field, as a result of damage by falling shade trees for instance, that further supplying should be done.

Development and yield

The rate at which cocoa trees develop depends on a number of factors; the more important being climate and its interrelation with soil, shade conditions and the planting material used.

Where the climate and soil allow continuous growth—conditions which are found in Malaysia—cocoa trees will form a jorquette within six to nine months of planting; the canopies will meet at a spacing of 3×3 m within eighteen months and the first crop may be gathered towards the end of the second year and certainly in the third year. Yields as high as 500 kg per hectare have been recorded in the third year. Such a rapid rate of development is unusual and would depend not only on suitable climate and soil, but also on light shade and freedom from pests and diseases. Under such circumstances the differences in rate of growth between various types of cocoa are greatly reduced.

At the other extreme, growth may be prevented during the wet and dry seasons, such conditions are found in India with its monsoon climate. The wet season is intense but lasts for only four to five months while the dry season is similarly intense and lasts for a similar period. The temperature falls with the onset of the monsoon, and above a certain altitude the fall may be sufficient to stop normal growth. The periods of growth are at the beginning and end of the wet season and may amount to only four to five months of the year. In most cocoa-growing countries conditions fall between these extremes. Growth will be reduced during the dry season but usually continues through the rest of the year.

Where the rate of growth is reduced by climate it will be influenced by the vigour of the planting material. The more vigorous Amazons and hybrids come into bearing earlier than Amelonado under Ghanaian conditions, but in Sabah there is less difference between Amazons and Amelonado in early yields.

After the first crops yields will increase for the following four to five years reaching a maximum eight to ten years after planting. The accompanying tables give examples of the way yields increase with age.

These figures show the way in which yields vary from year to year as well as increase with age. The large annual variations in yield are a feature of cocoa with which the grower has to become accustomed. In some instances, the change from one year to the next can be attributed to the presence or absence of a pest or disease or to measures taken to

Table 8.2 *Increase of yields with age (in kg dry beans per hectare)*

A. *Quoin Hill, Sabah*
   Two fields of 4·5 hectares each, planted in 1958.
   Shade: light jungle.

Field	Year								
	1961	1962	1963	1964	1965	1966	1967	1968	1969
3	630	695	922	898	884	1,652	1,305	1,490	1,262
5	381	358	447	288	750	1,140	984	1,140	1,010

SOURCE: *Ann. Rep. Dept. Agric. Sabah* 1967 and 1969.

B. *West Malaysia*
   Shade: thinned jungle.

Age of planting (years)	3	4	5	6	7	8	9	10
32 ha Amelonado planted 1954/55	—	—	153	218	481	643	724	802
8·5 ha Amazon planted 1956/57	60	320	490	623	997	1,488	1,228	1,009

SOURCE: Estate records.

C. *Cameroon*
   Commercial plantation.
   Shade: jungle.
   49 hectares F_3 Amazon cocoa planted 1958.

1961	1962	1963–64	1964–65	1965–66	1966–67	1967–68
117	234	430	736	838	996	600

SOURCE: Estate records.

D. *Trinidad*
   Clonal cocoa, Tortuga Estate.
   3·35 hectares planted 1943.

Age	3	4	5	6	7	8	9	10	11	12
Yield	173	328	822	834	1,020	884	1,078	1,350	1,270	1,880

SOURCE: Carib. Commission. Exch. Serv. No. 27, 1957.

control them; in others some extreme of weather may affect the crop adversely; but in many cases there is no obvious cause for the changes which take place. The minor changes in the weather pattern presumably have some effect on the crop but the importance of the various climatic conditions remains unknown so it is not possible to forecast crops

34. Cocoa farm in Ghana.

35. Cocoa farm in Ghana showing typical ground conditions in a healthy farm.

36. Amazon cocoa, five years old, on a plantation, West Cameroon.

37. Nacional cocoa, Ecuador.

38. (*left*) A cocoa field in Bahia, Brazil.

39. A typical small estate in Bahia.

40. A cocoa farmer's 'village' in Ghana.

41. (*right*) Regeneration by chupon.

from studies of the weather. This is an area of ignorance on which light may be shed in the future.

Having reached peak yields at seven to ten years, cocoa plantings are expected to maintain such yields on average until twenty years old, and thereafter to decline gradually. The yield of an area of cocoa is made up of two components; the individual tree yield and the number of bearing trees. There is some data on these factors, which give some indication of yield changes.

The detailed survey of cocoa estates in Trinidad, carried out by Shepherd (1937), provides unique data, showing that on good soils the yield of individual trees continues to increase up to seventy years. As there were no fields older than seventy years, it is possible that tree yields would increase beyond that age. On the other hand, tree losses, largely due to falling shade trees, led to a gradual decline in the yield of a field after thirty years. On poor soil the situation was different— both tree and field yields declined after twenty years, field yields at 6 per cent per annum.

In other countries similar yield patterns may prevail but the causes of death will differ. In Ghana the yield records for twenty-two years of random planted Amelonado cocoa were studied by Hall and Smith (1963). Initially the farms were about twenty years old and yielding 1,000 kg per hectare. During the following twenty-two years yields declined gradually to 500 kg per hectare and then increased to about 700 kg per hectare as a result of capsid control. The main cause of the gradual decline was virus disease which resulted in the loss of 42 per cent of the trees over the whole period. Without these tree losses, yields would have been maintained.

Judging by these two detailed studies of yields, a planting of cocoa can be expected to give good yields until twenty-five to thirty years of age, when replanting may be necessary. After that period better planting material ought to be available so that replanting should be worth while regardless of the likely decline in yield.

Rehabilitation

The rehabilitation of a cocoa field or estate means its restoration to a profitable position. Unprofitable plantings will have declined in yield and this first step is to determine the reason for low yield. The main reasons for decreasing yields are the increasing age of the trees, an unsuitable variety of cocoa or the diminishing productivity of site.

Diminishing productivity may be due to various factors:

1. Changes in the soil due to erosion, poor drainage or decreasing nutrient levels.
2. Competition from weeds and shade trees.

3. Too much or too little shade.
4. Attacks by pest or diseases.

Having determined the factors involved, the method of rehabilitation can be decided.

In some cases the cocoa farm can be restored by cultural means: the correction of shade, control of weeds, improving drainage, control of pests and diseases. At Bunso, Ghana, over 240 hectares of old Amelonado cocoa were restored by these means which raised yields from 280 kg per hectare to 960 kg per hectare in five years (Laryea, 1971). Similar results have been reported from Ivory Coast (Lanfranchi, 1971). These measures have been adopted fairly widely in West Africa for rehabilitation projects to deal with farms that have become moribund after capsid attack or have been neglected. They amount simply to good cultivation.

To many people rehabilitation will mean something more drastic, involving the renewal of all or most of the old trees. The term renewal is intended to cover replanting with seedlings or improvement of the trees by budding. The methods of replanting were the subject of lengthy debate in Trinidad where cocoa surveys had shown that 30 to 40 per cent of tree sites were either blank or held low yielders. When the Cocoa Subsidy Scheme was started in 1945 both partial and complete replanting were authorised. Partial replanting involves the removal of the poor-yielding trees and their replacement by new seedlings. Complete replanting involves removal of all the trees and replanting with shade and cocoa. Partial replanting was more popular because it did not involve any break in production from the replanted fields. Experience showed, however, that partial replanting is difficult to supervise and that its results are liable to be disappointing. Murray (1966) produced data comparing the two practices (Table 8.3) which shows that complete replanting will lead to much higher annual yields and will have surpassed partial replanting in annual and aggregate yield after five

Table 8.3   *Systems of replanting and yield of cocoa*

Year	Complete replanting	Partial replanting
	kg per hectare	
1	—	235
2	—	235
3	135	270
4	480	300
5	1,050	235
6	840	235
7	1,080	370
	3,585	1,880

SOURCE: Murray (1966).

years. Complete replanting is therefore recommended and it should be carried out field by field on old estates.

In Nigeria a rehabilitation experiment was laid down on a thirty-year-old farm. It compared the planting of F_3 Amazon seedlings under old cocoa with planting after clear-felling followed by various treatments. This form of underplanting produced the best results in terms of yield, percentage survival and growth; clear-felling followed by planting under *Tephrosia vogelii* gave the next best yield (Are, 1971).

When seedlings are planted under old cocoa there is a risk of disease —virus or vascular streak dieback, for instance—being transmitted to the seedlings. In these circumstances clear-felling would be better than planting under old trees. An alternative method is to coppice the old trees and bud the new chupons with better planting material. Trials of this method have been conducted in Nigeria (Are and Jacob, 1971). The trees are coppiced to a height of 30 cm and chupons will start to grow four or five weeks later. They can be budded when they are four to six months old, using patch-budding or some other budding technique. The rehabilitation of trees which are damaged or have fallen over has for long been done by allowing chupons to grow from the base and selecting the strongest. This did not allow any improvement in planting material but the chupons could be budded where better planting material is available.

The subject of rehabilitation is one to which attention has only recently been drawn. In Ghana and Nigeria there is little forest land left for cocoa planting so it is becoming increasingly important to rehabilitate the older cocoa areas, and a similar situation exists in Brazil. There is much to be learnt on the problems that arise in rehabilitating old cocoa areas.

References

Are, L. A. (1971) 'Cacao rehabilitation in Nigeria', *Proc. 3rd Internat. Cocoa Res. Conf. Accra 1969*, pp. 29–36.

Are, L. A. and Jacob, V. J. (1971) 'Rehabilitation of cacao with chupons from coppiced trees', *Proc. 3rd Internat. Cocoa Res. Conf. Accra 1969*, pp. 113–18.

Bonaparte, E. E. N. A. (1966) 'Pruning studies on Amazon and Amelonado cocoa in Ghana', *Trop. Agric.*, **43**, 25–34.

Brown, D. A. Ll. and Boateng, B. D. (1972a) 'Weed control in young cocoa: manual methods compared with a paraquat spraying technique', *Proc. 11th Br. Weed Control Conf. 1972*, pp. 466–71.

Brown, D. A. Ll. and Boateng, B. D. (1972b) 'Weed control for young cocoa: current work at the Cocoa Research Institute of Ghana', *Cocoa and Coconuts in Malaysia, Proc. Conf. Incorp. Soc. Planters, Kuala Lumpur, 1971*, pp. 145–54.

Evans, R. C. (1970) 'Preparation and maintenance weeding in immature/mature cocoa using herbicides', Kuala Lumpur, *The Planter*, **46**, 369–72.

Hall, T. H. R. and Smith, R. W. (1963) 'The performance of randomly planted West African Amelonado cocoa at Tafo from 1938 to 1960', *Ghana J. Sci.*, **3** (1), 35–43.

Jolly, A. L. (1955) 'The effect of age on cocoa yields', *Rep. Cocoa Conf. London 1955*, pp. 54–6.

Jones, T. A. and Maliphant, G. K. (1958) 'Yield variations in tree crop experiments with specific reference to cacao', *Nature, London,* **182**, 1613–14.

Kasasian, L. (1965) 'Chemical weed control in cocoa', *Cocoa Growers' Bull.,* **5**, 10–15.

Kasasian, L. and Donelan, A. F. (1965) 'The effect of herbicides on cocoa (*Theobroma cacao* L.)', *Trop. Agric.,* **42**, 217–22.

Lanfranchi, J. (1971) 'Régénération cacaoyère', *Proc. 3rd Internat. Cocoa Res. Conf. Accra 1969*, pp. 49–55.

Laryea, A. A. (1971) 'Cocoa rehabilitation in Ghana', *Proc. 3rd Internat. Cocoa Res. Conf. Accra 1969*, pp. 37–48.

Leach, J. R., Shepherd, R. and Turner, P. D. (1971) 'Underplanting coconuts with cacao in Malaya', *Proc. 3rd Internat. Cocoa Res. Conf. Accra 1969*, pp. 346–55.

Murray, D. B. (1966) 'Rehabilitation problems in Trinidad and Tobago', *2nd Sess. FAO Tech. Wkg party cocoa prodn. and protn,* Rome, 1966. Paper CA/66/5.

Newton, K. (1966) 'Methods of establishment and shade management', *Tech. Mtg. on Cocoa Prodn. Honiara 1966*, S. Pacific Comm., Paper 22.

Shepherd, C. Y. (1937) *The Cocoa Industry of Trinidad. Some economic aspects. Series III. An examination of the effects of soil types and age on yields*, Port-of-Spain, Trinidad, Govt. Printing Office.

Shepherd, R., Teoh, C. H. and Koo, K. M. (1971) 'Herbicide spraying techniques', *Crop Protection in Malaysia*, Incorp. Soc. Planters, Kuala Lumpur, pp. 6–37.

Swarbrick, J. T. (1965) 'Estate cocoa in Fernando Po', *Cocoa Growers' Bull.,* **4**, 14–19.

Walmsley, D. (1964) 'Irrigation and weed control', *Ann. Rep. Cacao Res. 1963*, Trinidad, pp. 53–5.

Chapter 9

Shade and Nutrition

D. B. Murray   *University of the West Indies, St Augustine, Trinidad*

In certain tropical crops, of which the most important are cocoa, coffee and tea, the normal practice is to grow the crop under the shade of taller trees. These shade trees may be specially planted at the same time as the cocoa or they may be trees left growing when the original forest is cleared for planting the crop.

The origin of the practice is lost in the past but it is reasonable to suppose that it arose at the time when early attempts were being made to grow the tree as a crop rather than merely to collect the pods from wild trees. It would probably have been found that it was difficult to establish the tree in full sunlight and that better growth was made when the young plants were shaded by taller growing trees. So by trial and error the traditional practice of the use of shade trees could have become established.

Why shade?

It is only in recent years that the function of shade in relation to the crop has become better understood and, more especially, the relationship of shade to nutrition which is the reason why these two factors are grouped together in this chapter.

The cocoa grower is interested in the answers to two questions: (1) What is the best degree of shade under which to grow the crop?; (2) What types and quantities of fertilisers are necessary to increase yield? It has been found that neither question can be answered without reference to the other, that is to say, the shade requirement and the nutritional or fertiliser requirement are interrelated and cannot be considered separately.

In dealing with the necessity or otherwise for the use of shade the argument has often been advanced that, since the natural habitat of the original uncultivated tree was the lower storey of the tropical rainforest of the Amazon basin, efforts should be made to reproduce this habitat in the commercial growing of the crop. Cocoa survives under such conditions but its yields are only such as to ensure the survival of the species. The cocoa grower on the other hand wants to grow his crop under conditions which ensure the greatest economic yield.

Effect of shade

The effect of natural or planted tree shade on the cocoa growing beneath must obviously be very complex. The reduction in light intensity is only one of several factors involved. Under tree shade both air and soil temperatures and their diurnal variation are reduced; wind movement is checked and atmospheric humidity and soil moisture are affected. Root competition must occur between the cocoa tree and the shade tree both for water and for mineral nutrients. If the shade tree is leguminous its root nodules may improve the nitrogen status of the soil, but this would not be so for other nutrients which would become locked up in the tissues of the tree. On the other hand, if the shade tree is deeper rooting than the cocoa its roots may exploit minerals from the deeper layers of the soil and these will enter the surface layer through leaf fall. Thus Adams and McKelvie (1955) report that on a typical shaded cocoa farm in West Africa, the forest tree shade contributed some five tons of leaf litter per hectare per year containing 79 kg nitrogen and 4·5 kg phosphorus. Finally dead roots from the shade trees may leave channels in compacted soils which will improve aeration and drainage.

A further important economic effect of shade concerns weed suppression. Fairly heavy shade in the establishment phase reduces growth of weeds, especially grasses, and the early closing of the cocoa canopy maintains this control.

Field experiments

Perhaps the earliest recorded field experiment to determine the optimum degree of shade for cocoa was started at River Estate, Trinidad, in 1910 by Freeman (1929). A field of forty-year-old cocoa was divided into three blocks. Each block had 500 trees spaced at 4·6 m by 4·6 m, that is about one hectare per block. In one block all the shade trees, *Erythrina* sp., were removed, in another, half the shade trees were removed and the third was left undisturbed. No special cultural or manurial treatments were applied. The yields of the trees were recorded up to 1928 and a summary of the results is given in Table 9.1.

Freeman concluded that the lightly shaded cocoa gave the highest yield and that as the crop was borne over a longer period it was easier to harvest. Both the lightly shaded and unshaded cocoa had a lower incidence of black pod disease. However as the plots were not replicated and moreover as their yields at the start of the experiment were

Table 9.1 *Shade experiment: yields, kg dry cocoa per hectare per year*

	Pre-treatment yield	Average for 18 years	Range between years
No shade	1,190	1,100	795–1,510
Half shade	1,110	1,230	910–1,760
Normal shade	940	930	480–1,220

markedly different these conclusions cannot be considered very precise. No information is available on the actual light intensities but heavy *Erythrina* shade probably transmits an average of some 30 to 40 per cent of the incident light.

Two points of interest may be noted: first, a yield of nearly 1,100 kg per hectare maintained over eighteen years without shade and without additional fertilisers, and secondly, a very large annual variation shown by the yields. This second feature complicates a great deal of field experimental work with cocoa, since uncontrolled soil and climatic factors can completely override the effects of experimental treatments.

Little further experimental work was done until Evans and Murray (1953) carried out an experiment in Trinidad using artificial shade. This removed the complicating factors of root competition and uneven shade distribution under shade trees. Bamboo slats were arranged above an experimental field to allow through 15, 25, 50, 75 and 100 per cent of full sunlight. Within each shade treatment, eight fertiliser treatments were applied consisting of the presence or absence of N as ammonium sulphate, P as superphosphate and K as muriate of potash in all combinations. The planting material was rooted cuttings of three ICS clones.

During the first twelve to eighteen months in the field growth measurements showed that heavy shade resulted in plants with a larger leaf area per plant than those lightly shaded. The greatest girths were found in the plants growing at the 25 and 50 per cent light levels, light intensities above and below this range resulting in smaller girths. At the low light levels the plants produced rather long spreading branches with large dark green leaves whereas at the high light levels growth tended towards a more bushy type of plant with small rather pale green leaves.

A number of temporary plants had been planted between the experimental plants so that they could be dug up at intervals and weighed. These weights permit a better measure of growth than girth, height, etc. The relative growth rates (RGR), that is, the percentage increase in dry weight per week for plants dug up at the end of the first year, are shown in Fig. 9.1. From this it is clear that the greatest growth in the first year had been made at a light intensity between 30 and 60 per cent of full daylight. At this stage fertilisers had produced little effect on growth, the light intensity being the overriding factor.

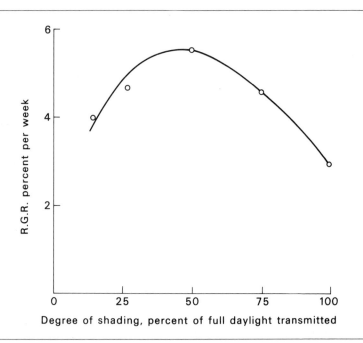

Fig. 9.1   Effect of shading on relative growth rate.

As the plants grew in size the effect of the fertiliser applications become increasingly evident depending, however, on the light intensity under which the plants were growing. Under heavy shade there was little difference between the treatments, but in the absence of shade the plants receiving the full NPK treatment were larger and greener than the controls receiving no fertiliser. When the plants came into bearing in their third year the effect of both shade and fertiliser treatment is best shown by reference to Fig. 9.2 which compares the yields from the plants receiving the NPK treatment with the unfertilised controls (Murray, 1954).

It is clear that at 15 and 25 per cent light intensities, that is, under heavy shade, yields are low irrespective of fertiliser application. With increasing light, yield increases up to 50 per cent light, but at light levels above this the yield is markedly affected by the presence or absence of fertiliser. In the absence of fertiliser yields fall off at light intensities greater than 50 per cent. With fertiliser, yield increases to its maximum at 75 per cent light. The practical significance of this finding is that, as stated earlier in this chapter, the shade requirement for cocoa and the response to fertilisers cannot be considered separately. They are interrelated, and this must be borne in mind when consideration is being given to the use of fertilisers on the crop.

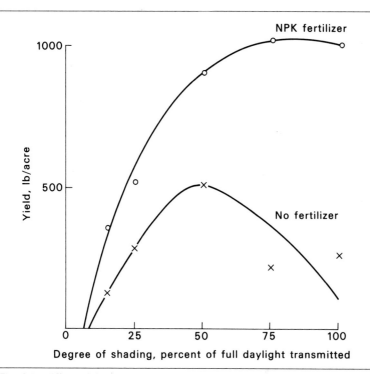

Fig. 9.2 Effect of fertiliser application on yield of cocoa grown at different light levels.

It is important to have some knowledge of the inherent fertility of the soil and its ability to supply the necessary nutrients for the plant. Thus on a highly fertile soil the conclusions arrived at above would indicate that optimal yield would be secured with little or no overhead shade. On a not very fertile soil, such as the one on which the experiment was carried out, a farmer would have two options. Either he could grow his cocoa under fairly heavy shade giving the cocoa about 50 per cent light and accept a not very high level of yield. Or he could reduce or even remove the shade, apply fertilisers and expect a far higher yield. Obviously this would need an examination of the economic value of the increased yield with respect to the cost of the added fertiliser. Confirmation of these results under artificial shade was given by a field trial at River Estate, Trinidad, and by an experiment in Ghana (Cunningham *et al.,* 1961).

The well-known trial at Tafo, Ghana, was laid down on a block of ten-year-old Amelonado cocoa planted at 2·4 m by 2·4 m under uniform shade of *Gliricidia* sp. The block was divided into several plots from half of which the shade was removed and it was further divided so that half the shaded cocoa was fertilised and half not fertilised. Thus there were four treatments with three replications. The fertilisers used were urea, superphosphate, sulphate of potash and sulphate of magnesium to give the following dosages:

| | *Rates of application (kg/ha)* | | | |
	N	P_2O_5	K_2O	MgO
First year	126	134	101	52
Subsequent years	112	112	84	56

Half the urea was applied in September, the other half with all the other fertilisers in April.

The effects on the yield of cocoa are clearly shown in Fig. 9.3. The absence of shade gave a substantial increase in yield which was further augmented by the fertiliser. Under the tree shade the effect of fertiliser was negligible. The practical implications are clear. Shade trees must be reduced in number, though not necessarily removed completely, to obtain the full benefit of fertiliser application. A warning is necessary on the danger of trying to increase yield by removing shade but not applying fertiliser. Depending upon the natural fertility of the soil, the increased drain on nutrients will lead in time to falling yields and early senescence of the trees. Field trials in countries as widely separated as Brazil, Ecuador, and Papua–New Guinea have shown similar results (Rosand *et al.*, 1971; Lainez, 1963; Anon, 1969).

Light intensity

In comparing results between different countries it is important to specify the actual radiation received. Incident radiation is affected by cloud cover, mist, dust, etc., which can reduce the total light intensity received by the cocoa even when unshaded.

Thus in Ecuador where the cocoa grows under heavy overcast conditions for most of the year, the average number of hours of sunlight per day as measured by a Campbell Stokes recorder is 2·6 hours, which is less than half the corresponding figure of 7·3 hours for Trinidad. Using the more relevant daily solar radiation measurements the corresponding figures are 294 Langleys for Ecuador and 390 for Trinidad. This might be interpreted to mean that unshaded cocoa in Ecuador is really partially shaded as compared with conditions in Trinidad.

In West Africa too, the overcast skies of July, August, September

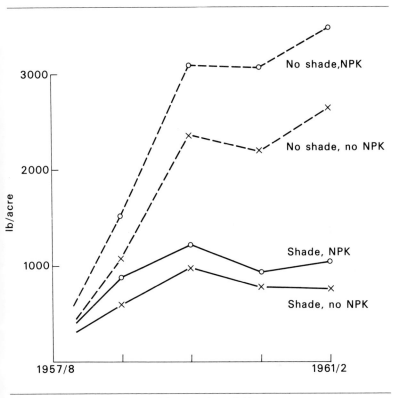

Fig. 9.3 Effect of fertiliser and light level on yield.

result in an average of 5·2 hours of sunshine per day at Tafo, a figure some 30 per cent lower than the 7·3 hours recorded in Trinidad.

Self shading

It must be realised that the relationship between shade and nutrition is not static. The optimum degree of shade will eventually be determined by the inherent fertility of the soil or by the addition of fertilisers, but temporary shade is always recommended during the establishment phase. It is possible to establish cocoa without shade but as demonstrated in Fig. 9.1 better early growth is made under shade.

The reason for the plant being able to tolerate more light as it grows is bound up with the concept of 'self shading' as the young plants grow in size. When first planted, the young seedlings or cuttings have only a few leaves each of which is almost fully exposed to the incident light. As the plant grows in size and produces more leaves the outer leaves tend to shade the inner ones so that the average light received per leaf is reduced. It follows that the light requirement for optimum growth and cropping increases as the plant grows in size and self shading becomes more marked. The final degree of self shading is approached when the canopies of the individual trees meet. The age at which this happens will depend on various factors including the spacing at which the trees were planted.

Types of shade tree

The general principle has been enunciated that the light requirement of the tree for optimum growth increases as the plant grows. When young however it does best under fairly heavy shade. In the field it is therefore necessary to have fairly dense shade established at the time of planting. This shade will already exist if planting is done under thinned forest, but if the area has been clear-felled temporary shade must be planted in advance of the cocoa. This temporary shade will vary from country to country, but one of the commonest is the banana or plantain. In West Africa other low-growing food crops are also used, such as the cocoyam, cassava and maize. The use of a shrubby leguminous plant has often been recommended, especially the pigeon pea *Cajanus cajan*, *Crotalaria* sp., *Tephrosia* sp. and *Leucaena leucocephala*.

As the cocoa grows this temporary shade is removed and gives place to the thinned forest tree shade or to the trees planted in advance of or at the same time as the cocoa. In the West Indies and South America this so-called permanent shade is invariably a leguminous tree such as *Erythrina* sp. (Immortelle), *Inga* sp. or *Gliricidia sepium*. In other countries it may be another economic crop such as the coconut, oil palm or rubber. In Trinidad, where a number of exotic trees have been tried as shade, nothing better than the Immortelles has been found.

Though there are obvious advantages to using an economic tree for shade, especially a timber tree, many of these compete with the cocoa and reduce yield. When thinned forest is used as shade local experience generally determines the trees which are compatible with the cocoa underneath and those which have an unfavourable effect.

Shade removal

One of the problems with shade trees is that with increasing age their crowns get larger and larger with a consequent reduction in the amount of light transmitted to the cocoa below. Murray (1955) showed that the yield of heavily shaded young cocoa could be doubled by removal of the shade, and this has been confirmed in West Africa. While the blanket removal of all shade is not a practice to be universally recommended the reduced light transmitted by the expanding shade trees is a factor in holding down yields of older cocoa at twenty-five to thirty years of age. Since the pruning of shade trees is hardly practical or economical the alternative is the removal of some of the shade trees.

One practice where Immortelles have been systematically planted at 7·3 by 7·3 m (14 by 14 ft) is to thin by poisoning at ten to fifteen years, first to 7·3 by 14·6 m (14 by 28 ft) and then a couple of years later to 14·6 by 14·6 m (28 by 28 ft). Shade trees should not be felled as they can do considerable damage to the cocoa beneath. With one of the recognised tree poisons, the trees fall in smaller sections which do less damage to the cocoa. This reduction in the shade tree population together with the application of fertilisers has led to substantial increases in yield in long-established cocoa fields in Brazil. Shade should be reduced slowly to allow the cocoa trees to adapt to the higher light intensity.

Shade or no shade

The preceding account of shade relations has explained the general theory of the shade–nutrition interaction. Stated simply, the better the mineral nutrition of the plant the less is shade necessary and the higher the yield. This must not be taken as a recommendation for the general and widespread removal of all shade. The relationship is an idealised one applicable when all other environmental factors are favourable. If this is not the case the relationship must be modified.

Factors which would militate against complete shade removal would include growth on a heavy clay soil with impeded drainage and a sharp dry season; under these conditions the rooting of cocoa is shallow and experimental work in Trinidad has shown that even with fertilisers shade is still necessary. In West Africa, mirid attack is more severe in the absence of shade. So, in the West Indies, is the incidence of the cocoa beetle, *Steirastoma breve* and of thrips, *Selenothrips rubrocinctus*. Little is known about the possible modification of the microclimate under shade in relation to the abundance of the pollinating insects. In consequenee it is recommended that shade reduction or removal should only be tried on a limited scale at first. Only after several years' experience should the practice be extended as it is far easier to remove shade than to replace it if that is found necessary. Moreover the widespread removal of shade trees over large areas of land could have marked effects on the general microclimate of the area.

Windbreaks

Before leaving the topic of shade, mention must be made of the importance of windbreaks in areas subject to drying winds. Cocoa is very sensitive to wind damage particularly in the dry season when availability of soil moisture is reduced. In areas near the sea it also suffers from a marginal scorching of the leaves, salt scorch, often thought to be a deficiency symptom.

It is considered that, in countries like Trinidad where the Immortelle shade tree is grown among the cocoa, part of its function is to act as a windbreak and to modify the microclimate of the cocoa below. In any case whether shade trees are used or not windbreaks are necessary wherever appreciable wind is experienced.

Windbreaks should obviously be sited at rightangles to the prevailing wind and their distance apart must be subject to local topography. A single species of tree may be used such as the mango but this tends to spread to such an extent as to overshade the neighbouring cocoa. With a less spreading tree such as the West Indian mahogany (*Swietenia mahogani*), galba (*Callophyllum antillanum*) or one of the *Eugenia* sp. it may be necessary to have a double storey with a lower growing plant as well. This can be *Dracaena* sp., *Hibiscus* sp. or *Aralia* sp. which are commonly used in the West Indies or any other type of hedge plant common to the country. Economic trees have obvious advantages but sometimes, as with the mahogany, root competition extends for two or three rows into the cocoa, with a resulting reduction in yield.

Mulch and organic manures

Although the prime purpose of a mulch is physical, to reduce evaporation loss from the soil and thus to conserve soil moisture, organic mulches also provide a source of nutrients and improve the crumb structure of the surface layer. Though it is difficult to separate all the possible effects of a mulch it is the intention to deal only with the nutritional effect.

The term mulch is rather vague and is usually taken to describe natural vegetation, be it grass or bushy material, cut and transported to the site to be mulched. Mulch in its widest sense, however, covers a range of materials from sawdust, bagasse, pod husks through composts to pen manure. Besides the differences in their physical structure and rate of breakdown the main distinguishing features in this range of materials lie in their nutrient content and its availability to the plant. Pen manure will obviously supply appreciable quantities of nutrients to the soil and thus to the plant, whereas the mineral nutrients available in a similar weight of sawdust or bagasse mulch will be much lower.

Before the advent of the relatively cheap artificial fertilisers, applications of pen manure and compost were the basic means of maintaining or improving the natural fertility of the soil. In the older cocoa-growing countries of Trinidad, Grenada and Sri Lanka the application of pen manure used to be a standard practice. Methods, rates and times of application varied so that the manure might be applied either by scattering on the surface, or digging in fairly uniformly, or in specially dug trenches between the trees or sometimes, as in Grenada, by tethering cattle among the cocoa trees.

In Grenada the standard practice was to apply a fairly heavy dressing of pen manure, approximately 50 tons per hectare, every four years and this was forked into the soil. Some depression in yield was usually found in the following crop due to root damage but this was more than compensated for by increased crops during the remainder of the cycle. In the cocoa-growing areas of West Africa the absence of farm animals due to the prevalence of trypanosomiasis precludes any use of pen manure.

The application of large quantities of bulky manure has become increasingly expensive, particularly as the terrain on a cocoa farm seldom allows the passage of any form of mechanical spreader. For economic reasons the practice has therefore to all intents and purposes died out.

The deterioration of cocoa soils under continuous cropping is a matter for serious concern particularly when the replanting of old cocoa is being attempted. Hardy (1960) has drawn attention in particular to the importance of a high level of organic matter and a high carbon/nitrogen ratio in the top 15 cm of cocoa soils to ensure high yields. The results of two experiments carried out at River Estate, Trinidad, are therefore of importance (Murray and Maliphant, 1964, 1965).

Tillage and manures experiment

The main treatments in this experiment compared the effect of (1) no surface cultivation—control; (2) shallow mechanical ploughing; (3) hand forking. Within each treatment there were four manurial treatments: (a) no manure—control; (b) pen manure fortified with NPK; (c) NPK only; (d) grass mulch fortified with NPK. The quantities of pen manure and mulch and their additions of fertiliser were adjusted so that the total amounts of nitrogen, phosphorus and potassium applied were the same for all treatments. This was done to separate the nutrient effects of the pen manure and mulch from such other physical effects they might have.

The average annual yields over a seven-year period are given in Table 9.2. It is clear that any form of soil cultivation reduces yield. This reduction in yield is mitigated by the application of manures but the obvious conclusion is that under River Estate conditions no form of soil cultivation should be practised. This is a general recommendation for tropical soils under high rainfall. Certainly with tree crops the less the surface layer is disturbed the better.

With regard to the effects of the manures, the plain fertiliser is

Table 9.2   *Tillage and manures experiment, River Estate, Trinidad: mean yields, kg dry cocoa per hectare, average for seven years*

	Uncultivated	Ploughed	Forked
Control	1,455	1,255	1,110
Pen manure	1,815	1,625	1,580
NPK	1,600	1,490	1,525
Mulch	1,890	1,825	1,770

responsible for an average increase of 145 kg of dry cocoa, the fortified pen manure for 358 kg and the fortified mulch for 440 kg per hectare per year. Deducting the increase due to nutrients only, i.e. the NPK treatment, it is seen that the pen manure and mulches were responsible for additional increases of 213 and 295 kg respectively. The conclusion would be that artificial fertiliser does not entirely replace the beneficial effects of the bulky organic manures.

Mulching and trenching experiment

This complementary experiment investigated the effect of trenching in which trenches 60 cm wide and 45 cm deep were dug between the lines of cocoa and filled with a mixture of pen manure and cut bush. This was compared with a surface mulch of cut bush and a bagasse mulch. Again additions of NPK were made to ensure that all treatments received the same quantities of these nutrients. The cocoa was grown without shade to avoid competition from shade tree roots.

Table 9.3   *Mulching and trenching experiment, River Estate, Trinidad: mean yields, kg dry cocoa per hectare, average for six years*

Control	900
Trenching	1,045
Bush mulch	1,273
Bagasse	1,060

As seen in Table 9.3, the treatment effects were disappointingly small. Even the increase with the bush mulch would almost entirely be accounted for by its content of mineral nutrients. Trenching led to a drop in yields the following year due to root damage. It must be concluded that bulky organic manures and mulches are not economic for cocoa but that every effort should be made to maintain and promote a high organic content in the top soil by natural leaf fall from the cocoa and shade trees.

Fertilisers

The point has been made that the fertiliser requirements of cocoa cannot be considered independently of the shade under which the trees are growing. Many of the early fertiliser experiments with cocoa were done under heavy shade and little or no response was found; in the case of additional nitrogen, generally given as ammonium sulphate, yields were actually depressed. It is essential therefore when considering the use of fertilisers that shade must be considerably reduced or removed altogether.

For cocoa established from virgin forest on fertile soils, fertilisers may not be required for many years but it is important to realise the export of the major nutrients from the soil that takes place in the crop each year. Figures for this export are given in Table 9.4.

Table 9.4 *Nutrients, kg, removed from the soil annually by a cocoa crop of 560 kg/ha*

	N	P	K
In beans	13	3·4	11
In husks	11	1·1	25
	24	4·5	36

Merely to replace the potassium removed by the world's crop requires some 60–70,000 tons of sulphate of potash annually.

Precise fertiliser recommendations are not possible without adequate knowledge of the soils of the area and their inherent nutrient content. Thus the first stage in determining fertiliser requirements is the soil survey and soil analysis. The interpretation of soil analysis has been discussed in Chapter 4 where the characteristics of soils suitable for cocoa are described. Soil analysis may reveal a gross deficiency or imbalance of nutrients which will assist in formulating fertiliser requirements. Leaf analysis may also assist in this respect.

Leaf analysis

Leaf analysis is a technique which aims at assessing the nutrient status of the plant from the quantities of the nutrient elements found in the leaf. It is usual to analyse only for the major nutrients, nitrogen, phosphorus, potassium, calcium and magnesium and the quantities found are expressed as a percentage of the dry matter in the leaf. By a comparison with standards established for optimum growth and cropping it should be possible theoretically to determine whether the level of a nutrient found in the leaf is suboptimal and to take steps to correct such imbalance by the use of fertilisers.

For some crops the technique has been developed to the stage where concrete recommendations for fertiliser usage can be recommended. With cocoa however, in spite of a great deal of research effort, the technique is only of limited value. Murray (1967) has summarised some of the problems involved.

To begin with, in taking leaf samples it is obvious that they must be comparable, i.e. be of the same age, come from the same part of the tree, etc. With the habit of leaf production by flushing the choice of leaves of the same age is not easy and the necessity for this is shown diagrammatically in Fig. 9.4. In the young fully expanded leaf the levels of N, P and K are high and they decrease with increasing age. On the other hand the levels of Ca and Mg increase with age. This means that unless the leaves being compared are of precisely the same age a low level of K in one leaf as compared with another may be due to the leaf being older and not because there is a deficient supply of K from the soil.

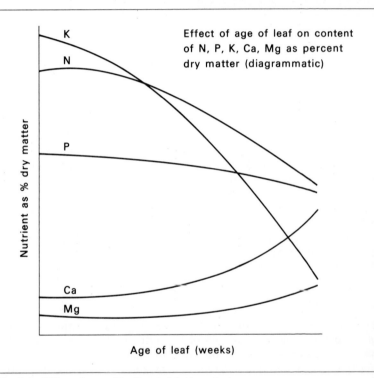

Fig. 9.4   Effect of age of leaf on nutrient content.

Another serious difficulty is associated with the light intensity under which the tree is growing. As shown in Fig. 9.5, trees growing under otherwise exactly comparable conditions will show higher levels of nutrients in their leaves under heavy shade as compared with trees growing under light shade or no shade. The implication of this is that so far as leaf analysis is concerned shade and mineral nutrition are

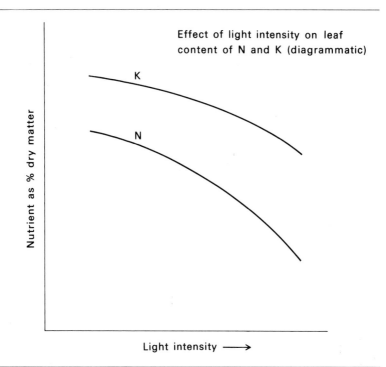

Fig. 9.5 Effect of light intensity on nutrient content.

equivalent. That is to say a low level of K in a tree growing without shade may be increased either by applying K as fertiliser or by shading the tree. The effect on yield would however be different, the use of fertiliser would be expected to increase yield while the increase of shade would decrease it.

Because of these problems the use of leaf analysis is only of limited value in planning a fertiliser programme. If a marked deficiency exists the low levels in the leaf will perhaps override the problems of sampling and indicate the type of fertiliser required to correct the deficiency. In the more normal range however the technique lacks precision. Table 9.5 from Murray (1966) gives the range of normal levels found in the leaf, low levels where growth and cropping may be affected but visual symptoms are not apparent, and deficient levels where visual symptoms are shown.

Visual symptoms of mineral malnutrition

By growing cocoa in sand and water cultures deficient in each of the nutrients in turn, the visual symptoms produced on the plant have been

Table 9.5 *Content of nutrients for normal cocoa leaves, leaves without definite deficiency symptoms and leaves showing deficiency symptoms*

Nutrient	Deficient	Low	Normal
		Percent dry matter	
N	<1·80	1·80–2·00	>2·00
P	<0·13	0·13–0·20	>0·20
K	<1·20	1·20–2·00	>2·00
Ca	<0·30	0·30–0·40	>0·40
Mg	<0·20	0·20–0·45	>0·45

described and illustrated by Maskell, Evans and Murray (1953) and Loué (1961). For full details of symptoms these papers should be consulted but a key follows. The first paper covers not only the effect of deficiencies but also of toxic levels of certain elements.

A. Symptoms more or less general on the whole plant.
   Element deficient—nitrogen, sulphur, phosphorus.
   Element toxic—boron.
B. Symptoms confined to, or at least more pronounced in, the older leaves.
   Element deficient—calcium, magnesium, potassium.
   Element toxic—aluminium, chlorine, iron.
C. Symptoms confined to, or more pronounced in, the younger leaves.
   Element deficient—iron, manganese, copper, zinc, boron, molybdenum, calcium.
   Element toxic—zinc, manganese, copper.

Symptoms are fully described in Appendix 3.

Field trials

The final and conclusive fertiliser recommendation should come from statistically laid out fertiliser trials in the field. These are expensive and time-consuming. They need to run over a period of years to allow for the physiological adaptation of the tree to the improved nutrition and for annual differences in rainfall and other environmental factors. It would not be possible to survey fertiliser trials in all cocoa-growing countries but, as an example of the type of experiment carried out, the results of a trial carried out in Nigeria by Wessel (1970a) follow.

The fertiliser rates used are given in Table 9.6 (*a*). No response was found to the potassium fertiliser but both nitrogen and phosphate applications gave increased yields as shown in Table 9.6 (*b*). It will be noted that a phosphate response was found in the absence of nitrogen, N_0, indicating that this was the primary deficiency. The nitrogen response occurred only in combination with the phosphorus.

Table 9.6 *(a) Fertiliser rates in kg per hectare per annum*

Level	N (as urea)	P_2O_5 (as triple superphosphate)	K_2O (as sulphate of potash)
0	—	—	—
1	90	34	56
2	180	67	112

(b) Yields in kg dry cocoa per hectare over five years

	P_0	P_1	P_2	*Means*
N_0	1,207	1,256	1,380	1,270
N_1	1,136	1,328	1,389	1,282
N_2	1,198	1,478	1,527	1,400
Means	1,180	1,354	1,422	1,317

From this and other experimental work a fertiliser mixture containing 20 per cent N and 20 per cent P_2O_5 was recommended, i.e. a 20:20:0 mixture at a rate of 375 kg per hectare. At a spacing of 3·7 by 3·7 m this would be the equivalent of 0·45 kg of fertiliser per tree per year.

He also showed (Wessel, 1970b) that young seedling cocoa benefits from monthly applications of 5 gm urea in the first year increased to 10 gm in the second year but warns of the dangers of overdosing the planting hole.

Obviously in another area where the soils contained higher levels of phosphorus but were deficient in potassium a formula would be needed containing potassium but lower in phosphorus as for example 10:5:15. From Ghana trials Cunningham (1963) recommends for unshaded cocoa, 125 kg urea, 125 kg triple superphosphate and 500 kg potassium sulphate per hectare to maintain yields of 2,240 kg per hectare.

In a more recent survey of fertiliser trials on shaded cocoa in Ghana, Ahenkorah and Akrofi (1971) have shown that the major response is to phosphorus with an optimum dose of between 45 and 90 kg P_2O_5 per hectare. They state that an economic return is secured only if the cocoa is already yielding at least 560 kg per hectare.

In general in full-grown cocoa where the canopy is meeting the fertiliser can be broadcast and it is not necessary to fork it in. It should be applied during the dry season so that it is not immediately washed away by rain. Splitting the dose to give two applications per year usually gives better results than a single application.

Minor elements

The diagnosis of minor element deficiencies is not always easy and they are usually first spotted from visual symptoms on the tree. Mention has already been made of lime-induced iron chlorosis. These symptoms are very easily recognised by the pattern of green veins standing out against a pale yellow green background. Although one per cent iron sulphate sprays can be used as a corrective this is not economic on a large scale and the basic solution is not to grow cocoa on soils approaching alkalinity. Zinc deficiency causes characteristic leaf symptoms, sometimes referred to as 'sickle leaf'; this deficiency can also occur under similar high pH conditions more especially where potassium levels are high, i.e. on sites where pods have been burnt in the past and the ash has accumulated. The characteristic symptoms of sickle leaf are unmistakable.

There is increasing evidence for the existence of boron deficiency in Ghana and in Ecuador. The correction of boron deficiency in Ecuador by foliar sprays of a soluble boron solution has been described by Mestanza and Lainez (1970). In Ecuador, sulphur deficiency also occurs. Where sulphur deficiency is found, nitrogen fertiliser should be supplied as sulphate of ammonia rather than urea or other non-sulphur containing fertiliser. Minor element deficiencies can be corrected by foliar sprays and soil applications but their confirmation and control needs specialist attention.

General fertiliser recommendations

Where cocoa is being planted on a fertile soil previously under forest fertiliser additions should not be needed. If, however, the land is being replanted from old cocoa and, more especially, if this has not been fertilised previously, fertilisers should be used for the young plants. A general purpose formula could be made up of ammonium sulphate 2 to 4 parts, single superphosphate 2 parts, sulphate of potash 1 part. Sulphate of potash is preferable to muriate since cocoa is sensitive to chloride toxicity. Alternatively, compound fertilisers of analysis 6:5:6, 10:5:10 or similar ratios could be used. In the first year in the field the young plant can be given 100 to 200 gm preferably in split applications. The amount of nitrogen is important; if the shade is dense N should be reduced. If it is too light, N should be increased for nitrogen and shade are negatively correlated. Thus if the plants look pale green through over-exposure to light the condition can be improved by additional nitrogen.

On soils known to be deficient in phosphate or having phosphate fixing properties, 200 gm of superphosphate should be mixed with the soil in the planting hole.

From the second to the fourth year in the field the dosage should gradually be increased to 0·6 to 1·0 kg per tree where spacing is 3·7 by 3·7 m. Adjustments can be made for closer spacings. At this stage the canopy should be meeting and most of the temporary shade will have been removed. If growing under light tree shade the mixture with the lower proportion of N should be used, if unshaded more N is needed.

If there is any indication of magnesium deficiency in the soil that element should also be included in the mixture. The potassium–magnesium balance in the soil is important for good production. Again if the soil is high in potassium this can be reduced or left out of the formula. No hard and fast rules can be laid down for all conditions.

Naturally fertilisers will only be of value if they increase the yield of cocoa appreciably and the economics of their use must depend upon local prices of fertilisers and the price secured for the crop. Fertilisers are unlikely to be economic unless basic yields are more than 560 kg per hectare and they increase yields by 220–340 kg per hectare. Their use also depends upon a high level of management including good weed control and disease and pest control. There is little point in producing more pods if they are lost to black pod. And as stressed right through this chapter, their effect is greatly reduced if applied to cocoa under too heavy shade. Maximum effect comes with much reduced or no shade.

References

Adams, S. N. and McKelvie, A. D. (1955) 'Environmental requirements of cocoa in the Gold Coast', *Rep. Cocoa Conf., London 1955*, pp. 22–7.

Ahenkorah, Y. and Akrofi, G. S. (1971) 'Recent results of fertilizer experiments on shaded cacao in Ghana', *Proc. 3rd Internat. Cocoa Res. Conf. Accra 1969*, pp. 65–78.

Anon (1969) 'Shade fertilizer trials', *Ann. Rep. (1966–67) Dept. of Agric.*, Papua and New Guinea.

Cunningham, R. K. (1963) 'What shade and fertilizers are needed for good cocoa production?', *Cocoa Growers' Bull.*, **1**, 11–16.

Cunningham, R. K., Smith, R. W. and Hurd, R. G. (1961) 'A cocoa shade and manurial experiment at the W. African Cocoa Research Institute, Ghana', *J. Hort. Sci.*, **36**, 116–25.

Evans, H. and Murray, D. B. (1953) 'A shade and fertilizer experiment on young cacao, 1', *Rep. on Cacao Res. 1945–51*, Trinidad, pp. 67–76.

Freeman, W. G. (1929) 'Cacao research', *Trop. Agric.*, **6**, 127–33.

Hardy, F. (ed.) (1960) *Cacao Manual*. Turrialba, Costa Rica, Inter-American Inst. Agric. Sciences.

Lainez, J. (1963) 'Fertilizing cacao under different ecological conditions', *Ann. Rep. 1963*, INIAP, Ecuador.

Loué, A. (1961) *Étude des carences et des déficiences minérales sur le cacaoyer*, Bull. 1. Institut Français de Café et du Cacao, Paris.

Maskell, E. J., Evans, H. and Murray, D. B. (1953) 'The symptoms of nutritional deficiencies in cacao produced in sand and water culture', *Rep. on Cacao Res. 1945–51*, Trinidad, pp. 53–64.

Mestanza, S. and Lainez, J. (1970) 'The correction of boron deficiency in Ecuador', *Trop. Agric.,* **47**, 57–61.

Murray, D. B. (1954) 'A shade and fertilizer experiment with cacao, III', *Rep. on Cacao Res. 1953*, Trinidad, pp. 30–7.

Murray, D. B. (1955) 'A shade and fertilizer experiment with cacao, IV', *Rep. on Cacao Res. 1954*, Trinidad, pp. 32–6.

Murray, D. B. (1966) 'Cacao nutrition', in N. F. Childers, ed., *Fruit Nutrition*, Rutgers, The State University, N.J.

Murray, D. B. (1967) 'Leaf analysis applied to cocoa', *Cocoa Growers' Bull.,* **9**, 25–31.

Murray, D. B. and Maliphant, G. K. (1964) 'Tillage and manures', *Ann. Rep. Cacao Res. 1963*, Trinidad, pp. 50–3.

Murray, D. B. and Maliphant, G. K. (1965) 'A mulching and trenching experiment', *Ann. Rep. Cacao Res. 1964*, Trinidad, pp. 35–40.

Murray, D. B. and Nichols, R. (1966) 'Light, shade and growth in some tropical plants', in *Light as an Ecological Factor*, Br. Ecol. Sym. No. 6, Blackwell Scientific Publications, pp. 249–60.

Rosand, P. C., Miranda, E. R. de and Prado, E. P. do (1971) 'Effect of shading and fertilizer application on yield of mature cacao in Bahia', *Proc. 3rd Internat. Cocoa Res. Conf. Accra, 1969*, pp. 328–38.

Wessel, M. (1970a) 'Fertilizer experiments on farmers' cocoa in South Western Nigeria', *Cocoa Growers' Bull.,* **15**, 22–7.

Wessel, M. (1970b) 'Effects of fertilizers on growth of young cacao', *Trop. Agric.,* **47**, 63–6.

Chapter 10

Diseases

R. A. Lass

It is difficult to estimate with any accuracy the losses of cocoa crops caused by diseases, but various attempts have been made. Hale (1953) estimated losses from both pests and diseases at 200,000 tons at a time when world production was 750,000 tons; this suggests that losses were about 21 per cent of potential production. Padwick (1959) estimated the losses from diseases to various crops grown in the then colonies; the loss to the cocoa crop was said to be 29·4 per cent, considerably higher than most other crops. More recently Cramer (1967) has estimated annual losses of cocoa from diseases to be 588,000 tons or 20·8 per cent of potential production when additional losses due to pests and weeds are included. Whatever may be the accuracy of these figures, they indicate a high rate of loss of crop from disease.

Attack by disease may result in a direct loss of crop as with black pod and other pod diseases, or the tree may be debilitated as with vascular streak dieback or even killed as by Ceratocystis wilt. To counter disease it is first necessary to identify the cause, then to assess the damage that is likely to be caused. Finally possible control measures should be examined and a decision taken as to whether they are worth applying.

Most diseases can be controlled by one means or another. This may involve simple field sanitation, which may include some change of shade or drainage; in other cases spraying will be necessary. Some diseases are more difficult to control, the more virulent forms of virus disease, for instance. In this case the diseased trees have to be cut down, but even so it is difficult to control or eradicate the disease.

It is preferable to try to prevent diseases and a most important aspect of prevention is through plant quarantine. Many plant diseases, including perhaps some affecting cocoa, have been transferred from country to country or even from continent to continent through ignorance or carelessness. This usually happens by the movement of vegetative material and the only safe way to move such material is through some intermediate quarantine station such as the Royal Botanic Gardens at Kew or the US Department of Agriculture Station at Miami, Florida. Seeds are comparatively safe; viruses are not known to be seedborne and the movement of fungal diseases by seeds could only occur through negligence or the omission of normal fungicidal treatment.

The aspects of importance to cocoa growers concerning the major diseases affecting cocoa are discussed in this chapter, but the finer aspects of pathology are not described in detail. Each disease of cocoa is considered under four headings:

Situation and outlook—assessment of the historical and future importance of the disease.

Symptoms—general outline to permit recognition of the disease in the field.

Epidemiology—important considerations in the spread of the disease.

Control—description of the recommended methods for reduction of inoculum level and spread of the pathogen.

Virus diseases

Following the discovery in 1938 that swollen shoot disease of cocoa in Ghana was caused by a virus, virus diseases have been found in several countries in West Africa and other parts of the world. The swollen shoot virus has caused enormous losses of crop and trees in Ghana and to a lesser extent in Nigeria. The virulent strains of this virus cause the death of Amelonado or sensitive cocoa trees in a relatively short time and the spread of such a disease naturally gave rise to great apprehension as to the future of cocoa-growing in Ghana. This situation led to an awareness of the possible danger from virus diseases in other countries; surveys were initiated and virus disease was found in Western Nigeria, Ivory Coast and Sierra Leone in West Africa and also in Trinidad and Sri Lanka (Thresh and Tinsley, 1958). Symptoms similar to those of virus diseases have been reported from Colombia and Venezuela, Java and Sabah but the presence of virus diseases has not been proved. Apart from West Africa, virus diseases have caused very little damage but the possibility of relatively harmless viruses causing more severe damage under different conditions is one reason for strict quarantine arrangements being applied to the movement of cocoa planting material. It should be emphasised that none of the cocoa viruses are known to be seedborne.

The relationship of the viruses found in Trinidad and Sri Lanka to those in Ghana is not known but more than one virus has been found in West Africa. While most of the virus disease occurring in West Africa is caused by isolates of the cocoa swollen shoot virus, two other viruses, cocoa mottle leaf virus and cocoa necrosis virus have been identified.

Cocoa necrosis virus was discovered in Nigeria and later in one locality in Ghana (Thresh, 1958; Owusu, 1971). This virus produces distinctive leaf symptoms of translucent distorted patches along the veins; shoots may wilt and die back but usually recover, although seedlings in Ghana have been killed. This virus is not transmitted by mealybugs.

Cocoa mottle leaf virus has also been found in Nigeria and Ghana, where it was originally found near Kpeve. It produces a red mottle on flush leaves followed by vein clearing and banding. The baobab, *Adansonia digitata*, is an alternative host of this virus.

Cocoa swollen shoot virus

Situation and outlook   Swollen shoot disease has been and still is a major problem for the cocoa industries of Ghana and Nigeria. It was first reported in 1936 when trees in the Eastern Region of Ghana were found to have developed stem swellings and dieback (Steven, 1936), but there is evidence that this condition was present in the New Juaben district as early as 1920 and probably since cocoa was first planted there in 1907 (Dale, 1962). In 1938 Posnette (1940) proved that the disease was due to a virus.

During the war years cocoa farms in Ghana tended to be neglected and the Department of Agriculture was only able to conduct limited surveys to find outbreaks of the disease. By the end of the war swollen shoot had killed the cocoa trees over a large area in the Eastern Region and production in that area had been halved. The policy of cutting out diseased trees on a wide scale was started in 1947 and, in spite of political difficulties in the early years and profound changes in policy and organisation in 1962, has been successful in containing the disease and saving the vast area of younger cocoa in Ashanti and Brong–Ahafo. On the other hand the disease was said to be out of control in the Eastern Region in 1971 (Kenten and Legg, 1971). The cutting out campaign involved a large organisation which carried out detailed surveys of the whole cocoa area of Ghana besides cutting out 140 million diseased trees between 1947 and 1969. Swollen shoot remains a menace and this work of surveying and identifying outbreaks of disease has to be continued.

In Nigeria swollen shoot outbreaks were discovered in 1944 in two areas, the larger being to the east of Ibadan. Control measures met considerable opposition and were abandoned in favour of trying to contain the disease with a *cordon sanitaire*. Trials carried out during the 1950s showed that the effect of virus disease alone was relatively slight and it was only where trees were attacked by capsids that virus infection accelerated the decline and possible death of the trees (Thresh, 1960). This finding led to the abandonment of the cutting out policy in Nigeria where it had never been implemented on the same scale as in Ghana. These trials raised the question as to whether cutting out was necessary in Ghana but it was clearly shown that the virus isolates found in parts of Ghana are more virulent than those in Nigeria and capable of killing trees.

Virus disease was found in Ivory Coast in 1943, at first close to the border with Ghana but later in the western cocoa areas. These outbreaks caused some concern initially but they have caused little damage and control measures are not in force.

Symptoms   There are many strains or isolates of the swollen shoot virus which differ in the symptoms they produce, the vectors that transmit them and their host range. The most virulent isolate is 1A or the New Juaben strain which can kill Amelonado seedlings in a few months and mature trees within two years. Most of the isolates found in western Ghana, Ivory Coast and Western Nigeria are far less virulent and some may have only a small effect on yield.

Swollen shoot virus 1A produces swellings on fan and chupon shoots and on the roots. The symptom caused by this isolate on young flush leaves is a red vein banding, followed by clearing or chlorosis alongside the veins. This is caused by pale chlorotic areas adjacent to the minor veins. At a later stage a fern leaf pattern is produced and mature trees in a chronic stage of infection have a generally yellowish appearance. Pods become mottled, smoother than normal and rounded, containing only half the normal weight of beans (Posnette, 1947). This isolate can reduce yield by 50 per cent in the first year after infection and may kill infected trees in the second year (Crowdy and Posnette, 1947).

Other isolates produce different symptoms and it is by the symptoms they produce that they are distinguished. Some isolates do not produce swellings; many isolates do not produce any pod symptoms. All isolates give rise to leaf symptoms which vary in detail, and with some virulent viruses have only been recorded on young experimental seedlings.

Epidemiology   Cocoa swollen shoot virus is transmitted from tree to tree by mealybugs. Over a dozen species of mealybugs are known to be capable of transmitting the virus, but two species, *Planococcoides njalensis* and *Planococcus citri* are the commonest. The common species *P. njalensis* is found in relatively small numbers mostly on young shoots in the canopy, although large colonies build up on pods. None of the mealybug species are found in sufficient numbers to be a pest in their own right and are important solely as vectors of virus diseases. *P. njalensis* and some other species are tended by ants which feed on the honeydew the mealybugs excrete and protect the mealybug colonies by building tents over them.

P. njalensis has a life cycle of about six weeks passing through three nymphal instar stages before becoming adult. All stages are mobile to some extent but the first instar nymphs, being numerically the most abundant, are more important in spreading the virus from diseased to healthy trees. The young mealybugs can become infective after feeding on a diseased tree for only one and a half hours, but infectivity increases with further feeding and they can remain infective for twenty-four to forty-eight hours after feeding ceases. The mealybugs are dispersed by crawling, by air currents, and possibly by their attendant ants; they can also be spread on pods after harvest.

The mealybugs are found on many other trees and plants in and around cocoa farms and a number of plant species have been found to be susceptible to swollen shoot virus. These species are fairly closely related to the cocoa tree and include *Cola chlamydantha*, an understorey tree occurring in the western part of Ghana, *C. gigantea*, *Ceiba pentandra*, the silk cotton tree, *Bombax buonopozense* and *Sterculia tragacantha*. With the exception of *C. chlamydantha*, it has proved difficult to transmit virus from these species to cocoa in the field.

Healthy cocoa trees which become infected may not show any symptoms for a considerable time. This latent period varies considerably according to the strain of virus and the age and condition of the tree. The virulent New Juaben strain may produce symptoms on mature trees within five months, but mild strains may not express themselves for two years or more. Infected but symptomless trees may be infective towards the end of the latent period. Where infection is suspected, as in trees surrounding an outbreak, the expression of symptoms can be accelerated by coppicing.

The spread of the disease may be discontinuous, or jump spread, probably due to mealybugs being carried some distance by air currents. Generally, however, the spread is gradual, a slow but steady process by which an outbreak becomes progressively larger causing ever increasing numbers of trees to become infected.

Control Virus infections in plants cannot be cured; the only way to restrict spread is to destroy the virus sources. With cocoa swollen shoot, the removal of infected trees has slowed down the spread and occasionally eliminated outbreaks, but to do this is an enormous and difficult task involving the locating of scattered infections in some 4 million acres (1·6 million ha) of cocoa in Ghana.

In Ghana the practice has been to cut out only those trees with symptoms of swollen shoot. Cutting out involves cutting the trunk below ground level so that infected trees cannot regenerate. After initial treatment, outbreaks are reinspected and retreated on a regular basis, the reinspections being continued until no infected trees are found for a period of two years. This method is fairly effective in controlling small outbreaks but with large outbreaks there are so many latent infections that the method is ineffective. The cutting out of two rows of contact trees would be much more effective but the practice is not acceptable to farmers.

In Nigeria the original policy was to cut out all trees within 30 m of an outbreak, a rather drastic treatment but one that was effective. This was revised in favour of removing symptomless trees over smaller radii depending on the size of the outbreak. Later still the cutting out campaign in Nigeria was abandoned.

Attempts have been made to control the mealybug vectors by the use of insecticides and by biological control, but the results have been unsuccessful. An effective insecticide was found but it was highly toxic and therefore dangerous to apply and it produced a taint in the cocoa beans. The possibility of controlling mealybugs by eliminating their attendant ants has been tried but the persistent pesticides used were only partially effective and led to disastrous side effects, pod and stem borers becoming important pests (Entwistle *et al.*, 1959).

In the long term the disease may be controlled by the use of resistant varieties, and a great deal of research has been devoted to this end. The Amelonado variety is uniformly susceptible to swollen shoot virus but Amazon types exhibit variation in susceptibility. In the first place interest was centred on the possibility of using tolerant varieties, i.e. those which yield well in spite of infection, but results were unsatisfactory and this approach has been given up in favour of selecting for resistance. This is a long-term programme requiring a great deal of testing of selected varieties, but it should be ultimately successful (*see* Legg, 1972). There is evidence that cocoa varieties influence the mealybug population (Bigger, 1972) so that changes in farming practice involving new varieties coupled with reduced shade may improve the virus disease situation.

Diseases primarily affecting flower cushions and pods

Black pod (*Phytophthora palmivora*)

Situation and outlook   Black pod disease is caused by the fungus *Phytophthora palmivora* which occurs in all cocoa-growing areas. Global losses due to this disease are enormous and were estimated by Padwick (1959) at 10 per cent of world production, a figure which is generally considered to be conservative. There is a wide diversity of incidence of black pod infection from almost zero in Malaysia to 95 per cent in the wetter parts of the Cameroon Republic (Tollenaar, 1958). Black pod incidence of 25 to 30 per cent has been recorded in Ghana (Wharton, 1962), 75 to 90 per cent in Nigeria (Gorenz, 1972) and 25 per cent in Brazil (Miranda and da Cruz, 1953). It is thus of enormous significance in some areas and may even determine whether it is profitable to plant cocoa. Chemical control by spraying with copper fungicide is the well established method of control but is expensive and not completely effective. Cultural techniques, crop sanitation and suitable management may reduce infection, but black pod infections will inevitably be present if the area is subjected to prolonged periods of high humidity.

A five year survey in Ghana found a significant positive correlation between the number of black pods and the total number of pods per acre (Blencowe and Wharton, 1961). The extent to which this relationship holds true may vary, but there will always be a greater absolute loss at high yields. The black pod problem will therefore become increasingly severe as higher yielding material is planted, and this will become more serious if the new planting material is more susceptible than existing types.

The pathogen can also infect flower cushions, leaves, shoots, seedlings and roots, and in addition a cocoa canker caused by *Phytophthora palmivora* occurs in many parts of the world. Canker caused by this pathogen was the most important disease early in the twentieth century, when plantings were based on Criollo selections. The Forastero types planted subsequently, particularly the Amelonado of West Africa, have considerable resistance to canker and this type of infection was generally forgotten. However, some of the Amazon types are susceptible and the disease has resumed some of its former importance (Vernon, 1971). While cankers can damage trees severely, it is difficult to assess this damage in economic terms.

Symptoms   On pods, the first sign of infection is noted about two days after initial infection, when a minute translucent spot will appear on the pod surface. This spot soon turns to a chocolate brown colour, then blackens and expands rapidly so that the whole surface of the pod can be blackened within fourteen days. The fungus produces spores from the infected area three days after initial infection to give a white or yellowish down over the already blackening area on the pod surface. Internally, the beans become infected about fifteen days after the initial infection and are soon commercially valueless. In pods approaching maturity the beans become separated from the pod wall and may escape infection; thus the beans can often be saved from ripe pods attacked by this fungus. Infected pods left on the tree remain a source of infection until they become covered with saprophytic moulds and cease to be a source of inoculum.

Phytophthora cankers can arise from direct contact with a diseased pod left on the tree or from an extension of a pod infection, when the fungus ascends the stem invading the flower cushion and then the surrounding tissues. Newly infected bark shows no external symptoms, but a pink-red discoloration will be found beneath the bark. The subsequent development of the infection depends on the susceptibility of the tree. In resistant cocoa the spread is halted and host scar tissue forms around the lesion and secondary saprophytic fungi take over completely. The infected tissue then turns brown and is invaded by insects, which clear away the diseased tissue leaving a clean scar. On susceptible varieties, however, the active phase is more prolonged and the secondary fungi do not take over. It seems that such a canker can remain in an active state for several months, possibly years, and will act as a reservoir of infection (Vernon, 1971).

Infection of other tissues can also occur but are not normally severe. When a chupon is attacked it generally dies back from the tip. The point of attack is usually in the axil of a leaf, although infection can first take place on a leaf-blade and spread backwards. At first the affected area appears water-soaked, but it soon darkens in colour and becomes sunken. The stem will be girdled by the infection and die above the lesion (Briton-Jones, 1934). Leaf infections developing from the tip have been noted throughout West Africa as a wet-rot moving rapidly down the main veins. Older diseased leaves may exhibit a dark brown withered tip which sometimes falls off (Turner and Wharton, 1960). Chant (1957) describes an infection of seedlings in nurseries, which killed 70 per cent of seedlings in one nursery under conditions of high humidity. The first symptom noted is a brown discoloration of leaves at the tip of the seedling which may lead to infection of the cotyledons and eventual collapse of the stem. If the cotyledons do not become infected, the seedling may recover due to development of axillary buds. Research in Ghana has shown that *Phytophthora palmivora* is also a root patho-gen of cocoa (Turner and Asomaning, 1962). Root infections could account for poor establishment in some areas when seed is planted and may also act as a source of inoculum for future infections.

The name *Phytophthora palmivora* is applied to a group of closely allied fungi which attack many plants including rubber, citrus, and coconuts. The relationship between the fungi and the possibilities of cross infection between the crops is far from clear, but the fungus attacking cocoa has been divided into two mating types A1 and A2, or the 'rubber' and 'cocoa' types respectively. The predominant form is A2 and Turner (1960) found that in West Africa the cocoa type occurs in Ghana and Ivory Coast, while the rubber type is found in Nigeria and Cameroon.

Epidemiology   The fungus can exist in several phases and will pass from one stage into another, dependent on conditions (Gregory, 1969), though the latter have not yet been clearly defined. The fungus is dis-persed by the sporangia, which are freely formed over the surface of pods at humidities of 60 to 80 per cent and temperatures of 20 to 30°C to give the characteristic white down. The conditions for dispersal are also not known, but spores may germinate to produce mycelia, zoo-spores or more sporangia. The simplest form of development is the growth of a new mycelium from a spore, the mycelium growing actively and causing the damage to pods and other invaded tissues. Where there is a film of water, the sporangia will divide into as many as

42. Effect of shade on growth: **(a)** light shade giving compact growth; **(b)** heavy shade giving spreading growth.

43. Leaf symptoms of swollen shoot virus disease *(from left to right)*: **(a)** vein-banding on a young flush; **(b)** angular chlorotic spots bordering veins; **(c)** interveinal bands; **(d)** "pepper and salt" mosaic; **(e)** red vein-banding.

44. A swollen shoot outbreak after cutting out with stumps stacked for checking.

45. (*right*) Black pod disease, Grenada.

46. (*far right*) Black pod disease, Cameroon.

47. (*left*) Spraying to control black pod with pressurised knapsack sprayer.

48. Monilia disease

49. (*left*) Monilia disease, Ecuador.

50. Witches' broom disease showing brooms and malformed pods.

51. Witches' broom disease.

52. Green point gall.

53. (*left*) Flowery cushion gall.

54. Dieback in Cameroon, probably caused by lack of shade.

55. Thread blight caused by *Marasmius*. The mycelium can be seen on the stem and undersurface of a dead leaf. The dead leaves are typically bicoloured.

56. (*right*) Unhealthy cocoa tree with heavy growth of mistletoes.

thirty zoospores which can move by means of two flagella. Each zoospore can settle and produce a mycelium. Mycelium within pod tissue may also develop a type of thick-walled spore to survive an especially unfavourable period. The pathogen is also known to be capable of sexual reproduction and thus of genetic mutation, but sexual spores have only been noted in culture media.

The spores of the pathogen require conditions of high humidity for germination and therefore it enters a resting stage during the dry season in all cocoa producing countries. Tarjot (1967) studied an old Amelonado plantation during the dry season and found *P. palmivora* in quantity on debris on the ground, in the soil, in the flower cushions from ground level to a height of over 2 metres, in the bark of the trunk and branches, on stalks of pods, on blackened cherelles and pods left on the tree, on pod heaps and on the bark of shade trees. Any one of these potential sources of inoculum can form a reservoir of infection at the start of the rainy season. Dade's work in Ghana (1928) showed that flower cushions are an important source of inoculum at the onset of an outbreak; in Nigeria there is evidence that the soil is the most dangerous source of infection (Okaisabor, in press) while in Costa Rica infection of leaves in the canopy is thought to be the initial source of infection (Manço, 1966). There are, therefore, many sources of inoculum and the relative importance of each source differs from country to country and possibly for different types of planting material. These differences, on which there is still much to be learnt, will influence the control measures to be taken.

In most cocoa-growing countries the empty husks of both healthy and infected pods are left on the ground after opening. The infection of healthy pods from diseased husks seems obvious; extension services in Cameroon and Brazil recommend the removal or spraying of diseased husks, and at one time the burial of diseased pods was advocated in Nigeria. The cost of these measures would be considerable but their effectiveness in controlling the disease has not been proved. While diseased pod husks will continue to liberate spores after harvesting they are rapidly invaded by harmless saprophytic moulds which dominate the husk after some time. It seems unlikely, therefore, that husk piles form an important source of inoculum.

The development of mycelium from a diseased flower cushion into a developing pod is one way in which an outbreak of black pod can be started. Infection from other sources infers the movement of spores within the cocoa farm. Attempts to trap spores in the air have been generally unsuccessful so that wind dispersal does not seem to be of any importance. Rain splash and various insects are evidently the main means by which the spores are spread. In Nigeria it has been shown that rain splash can carry spores from the ground to a height of one metre; leaf litter acts as a barrier between inoculum in the soil and healthy pods and delays the onset of an outbreak considerably (Okaisabor, 1972).

Evans (1973) has shown that insects are involved in the spread of black pod disease in Ghana. In particular the small black ant *Crematogaster striatula* uses dead plant tissues including infected and old black pods to construct tents around the pod peduncle and this can lead to the development of infection from the peduncle region. Once the soil is contaminated there will therefore be a continual re-cycling of inoculum downwards via rain splash and upwards via ants. These ants are probably important early in the season in creating new infections. Horizontal movement of spores may be accomplished by flying insects which feed on diseased pods and carry the spores internally or sticking to their mouthparts.

Little work has been completed on spore movement in other pro-ducing countries, but similar processes may be involved. Once an out-break of black pod disease has started, diseased pods will form the most important source of inoculum and the disease spreads largely by rain splash, the rain droplets carrying spores from diseased pods to healthy ones. At this stage the disease is very difficult to control as newly infected pods produce spores within two or three days of the first symptoms of attack.

Germ tubes from zoospores or the sporangia can penetrate the pod surface at any point and at any age of pod to give a mycelium, which infects the husk tissues provided the humidity is very high or there is free water on the pod surface. Climatic conditions are obviously important in initiating an epidemic and different theories have been put forward as to the relevant factors. Work in Brazil (Lellis, 1952) and in Sri Lanka (Orellana and Som, 1957) indicate a relationship between temperature and disease incidence. On the other hand, detailed studies of relative humidity in West Africa (Dade, 1927; Wood, 1974) have shown that long periods at saturation point are necessary for the rapid spread of the disease. Similar findings have been reported from Ivory Coast (Tarjot, 1971). The theory that relative humidity is the most important climatic factor helps to explain the higher incidence in Nigeria than in Ghana and the almost complete absence of black pod disease in Malaysia. Dade (1927) is considered to be right in his conclusion that 'atmospheric humidity is probably the most important factor affecting the incidence of black pod disease'.

In much of West Africa black pod disease makes its appearance a few weeks after the start of the wet season in April, but the incidence of the disease remains low until July when the rains are heavier, the skies generally overcast and humidity is high. The disease is normally halted by the short dry season, which occurs in August but incidence rises again when the rains resume in September. The sunnier weather which starts at the end of October brings the outbreak to a quick conclusion. In West Cameroon and in occasional years in other parts of West Africa, there is no short dry season and this results in much more severe losses during August and September unless control measures are effective.

Control From flowering to picking each pod can be at risk for five months and pods are present on the tree for most of the year, although conditions do not necessarily favour spread of the pathogen throughout this time. Spraying with a copper fungicide is the standard control measure, but is expensive and never completely effective. The use of other chemicals is possible and cultural techniques can reduce the level of infection. The long-term solution must lie in breeding for resistance or tolerance to the disease and in obtaining a better understanding of the method of spread. There has been considerable research effort in many countries on breeding, but no selections offering a high degree of resistance have been developed so far.

Breeding for resistance Field observations on cocoa varieties show consistent differences in incidence of black pod, though no selections have shown complete immunity. The replacement of susceptible trees by material showing durable resistance to the pathogen is the ultimate long-term solution to black pod losses.

Not all breeding programmes are currently considering black pod resistance as a criterion for selection, standard screening tests for resistance are still under development and the pathogen itself is an assemblage of strains differing genetically and is liable to variation (Gregory, 1972). These factors, combined with the long generation time of cocoa, mean that material with durable resistance is not likely to be planted for many years. The tendency for new plantings to be of Amazon hybrid material, which produces pods throughout the year is likely to increase the level of inoculum, because small numbers of pods may be present on the trees for most of the year. A promising approach to severe losses from black pod infection is to select for 'black pod escape', which infers that the tree produces the bulk of its crop when climatic conditions are least conducive to the spread of the disease (Atanda, 1973).

Cultural techniques Control of black pod can probably be assisted by certain cultural techniques, of which regular harvesting of ripe pods if often recommended. In view of the rapidity with which a newly infected pod starts to produce spores, it would be necessary to harvest or inspect every other day in order to control the disease. This treatment reduced the rate of infection from 90 per cent to 60 per cent in Nigeria (Thorold, 1959), but the cost involved makes it an impractical proposal for any area. It is, however, essential for various reasons that the pods are harvested regularly and that all diseased pods should be removed; this is particularly important prior to the wet season.

The effects of fertiliser, spacing and shade on black pod incidence are interrelated. Shade trees tend to prevent air movement in the cocoa canopy and shaded pods are likely to remain wet, and therefore at risk, longer than an unshaded pod. Vernon (1966) discusses this relationship and suggests that it is only relevant at higher yields. The current tendency in many cocoa-growing areas to reduce shade, apply fertiliser and achieve higher yields, may also reduce the proportion lost to black pod. Moreover shade removal may increase yields to a level at which chemical control may become economically viable (Gregory, 1969). Regular pruning, removal of diseased pods at harvest, avoidance of large heaps of pod husks and regular weeding to keep humidity in the farm as low as possible, probably all help to reduce the level of inoculum. There is, however, no experimental evidence to prove any of these suggestions and thus it is difficult to make recommendations concerning these practices.

Chemical control   The chemical control of black pod depends on copper-based fungicides, but throughout the world there is great variation as to the volume and concentration of the spray, the timing and frequency of application, the parts of the tree sprayed, the machines used and the fungicide itself. No detailed recommendations can be given, but it must be emphasised that fungicide spraying is a prophylactic measure protecting the pods against infection. The first spray must therefore be applied before the inoculum can build up and spraying must be repeated regularly as the effectiveness of the protective layer is continually being reduced by the action of the weather and by pod expansion.

In Nigeria, fungicide spraying started in 1953 following experiments using one per cent carbide Bordeaux mixture, applied with pressurised knapsack sprayers. The volume varied with the crop from 200 to 500 litres per hectare and averaged 315 l per ha. The sprayed area of 16 ha had a black pod percentage of 6·2 per cent whereas losses amounted to 65 per cent on 5 ha on which the trees had been inspected and black pods removed every other day (Thorold, 1953a). The success of this trial lead to widespread spraying by farmers. The use of carbide Bordeaux (0·45 kg copper sulphate, 0·17 kg calcium carbide in 45 litres water) has been replaced by 1 per cent lime Bordeaux and several proprietary copper fungicides, generally based on cuprous oxide.

The work in Nigeria was far from being the first attempt to control black pod with copper, experiments having been conducted in Cameroon as early as 1907. The practice of spraying with copper has been long established in Fernando Po where very simple syringe-type pumps are used to apply 1 to 2 per cent Bordeaux mixture at high volume—1,000 to 2,000 l per ha. Only two or three applications are made each year at low altitudes but higher up a stronger solution is applied more frequently (Swarbrick, 1965).

On an estate in West Cameroon spraying started in April or May and was continued on a three to four week cycle until the end of September. Simple knapsack sprayers were used to apply a proprietary cuprous oxide fungicide at 6·2 kg per ha for the first cycle and 4·2 kg per ha for subsequent cycles; the same mixture was used in all cycles, the first used 540 l per ha, later cycles about 340 l per ha, the volume varying according to the crop (Wood, 1969).

In these examples 1 per cent Bordeaux mixtures and similar strengths of other fungicides were used. Trials in East and West Cameroon indicated that stronger solutions were needed in wetter areas such as West Cameroon (Muller and Njomou, 1970).

It is clearly important to start spraying before the disease starts to build up and the vital period would seem to be the weeks or months before an epidemic starts. If the disease can be controlled at this stage then the losses may be contained better when the weather makes spraying more difficult. Spraying at three-weekly intervals has been generally recommended and found to be the most satisfactory interval in various trials.

As the principal object in spraying is to cover the pods with a protective coating, spraying is often confined to the pods. In Central America the canopy is often sprayed as well because lesions caused by *P. palmivora* are thought to be an important source of inoculum. In Fernando Po very large volumes of fungicides are used and much falls on the ground, which may help in controlling the disease. In an experiment in Nigeria comparing the effect of fungicides applied to the ground and lower part of the trunks two fungicides, captafol and Perecol, were effective in delaying the appearance of the disease (Okaisabor, 1972), but more work will be needed before the value of this can be assessed.

Various types of spraying machines have been used and a comparison of mist-blowers and knapsack sprayers was conducted in Nigeria (Hislop, 1963). Mist-blowers using low volume and less labour might appear to be preferable to knapsack sprayers, but as the mist cannot be directed at pods a larger quantity of chemical has to be applied and the cover is less effective. The simpler type of knapsack sprayer is generally found to be the most effective. It is essential that the task of spraying be closely supervised to ensure that the pods are being adequately protected (Filani, 1972).

In trials of fungicides Bordeaux mixture has generally proved to be slightly more effective than other copper fungicides. It is, however, inconvenient to mix and highly corrosive so that the proprietary copper fungicides are usually preferred. These fungicides are based on cuprous oxide, copper oxychloride or copper hydroxide and there are also 'instant' formulations of Bordeaux and Burgundy mixtures.

In addition to copper fungicides, two compounds, triphenyl tin chloride and triphenyl tin hydroxide, have been found to control black pod disease in East Cameroon and in Nigeria (Braudeau and Muller, 1971; Weststeijn, 1968); other compounds have shown promise in laboratory tests. No systemic fungicide has been found to be effective against black pod disease.

In Nigeria the addition of DDT to the fungicide improved the control of disease (Weststeijn, 1968).

Chemical spraying is expensive and farmers have to decide whether it will be profitable. Newhall (1966) produced tables showing how profitability of chemical control changes with the cost of spraying, the price received for cocoa and the expected yield increase. It might seem possible for every farmer to give consideration to the external factors at the start of each season and make a decision as to whether to spray against black pod. There is however a cumulative effect of spraying which builds up over the first few seasons from commencement of spraying and this cumulative effect will be lost if spraying ceases for a short period.

Monilia pod rot

Situation and outlook   This disease is also called watery pod rot or Quevedo disease after the area in Ecuador where it caused severe losses after its appearance about 1914 (Rorer, 1918). It is thought (Ampuero, 1967) that the disease originated in Ecuador and spread south to Peru and north to the cocoa areas of Colombia outside the Amazon basin. It also occurs in the most westerly cocoa area in Venezuela and in the southern part of the Panama isthmus, but has not been reported elsewhere in Latin America or any other cocoa producing country. Monilia pod rot and witches' broom disease were the major causes of the decline in production in Ecuador from nearly 50,000 tons in 1915 to 15,000 tons in 1930.

The disease is caused by the fungus *Monilia roreri* which only attacks cocoa pods and for which no alternate hosts have been found. In Ecuador and Colombia, where the disease is serious, it causes the destruction of 15–80 per cent of the pods. The magnitude of the loss seems to be correlated with climatic conditions at the time of pod-setting.

Symptoms   The pods are infected when they are young and the infection develops internally as the pod grows. The first indication of the disease is the appearance of spots of mature coloration on the surface of immature pods with no other external symptoms. These spots turn brown and later enlarge rapidly in size to cover the entire surface of the pods. Under favourable weather conditions the spots become covered with a layer of white mycelium bearing abundant spores. These spores can remain attached to the surface of the pod for a long time as a fine powder and thus act as an easily dispersable source of inoculum (Jorgensen, 1970). Some infected pods may not have any external symptoms, but when such pods are opened abundant liquid will be present, resulting from tissue degeneration. Pods infected with *Monilia* are heavier than healthy pods of the same age. The beans in infected pods may be partially or completely destroyed depending on the stage of maturity at which infection occurs.

Epidemiology The fungus remains as an asexual spore in old diseased pods between the wet and dry season. Spores are disseminated by wind, insects and rain, and it is believed that wind is the most important means of spread in Ecuador. Some workers found it impossible to inoculate mature pods and a correlation has been found between rainfall during the early stages of pod development and disease incidence (Desrosiers *et al.*, 1955). This suggests that only young pods are susceptible. Pods damaged by insects, particularly the stink bug *Antiteuchus (Mecistorhinus) tripterus*, are more liable to become infected but the fungus can infect undamaged pods (Franco, 1958). The first symptoms of the disease appear two months after inoculation.

Control Favourable conditions for *Monilia* infection occur throughout the rainy season in Ecuador. This is a time of considerable flowering and fruit setting, and is thus a critical period for protection of young pods. In Ecuador various chemicals, including copper and sulphur, have given some control when sprayed on a 10 to 14 day cycle during the rainy season, at high or low volumes (Ampuero, 1967). It is possible to restrict spraying cycles to the period of peak flowering and fruit setting, when conditions are favourable for infection. Jorgensen (1970) states that spray trials have given poor results and concludes that it does not pay to spray in old plantations and that it is very doubtful whether it pays to spray in the younger plantations. In a year of low prices or low infection spraying is almost certainly not economic.

In Ecuador the Trinitario or Venezuelan types suffered greater losses than Nacional trees and there should be a basis for selecting resistant types but no selections have so far been tested. Until resistant varieties can be developed, harvesting at frequent intervals and destruction of infected pods are measures which will help to reduce the level of primary inoculum.

Witches' broom

Situation and outlook Witches' broom disease of cocoa is a factor limiting production in several countries of the Western Hemisphere. It is caused by the fungus *Marasmius perniciosus*, which was identified in Surinam (Stahel, 1915), where the disease first appeared in 1895 and caused a rapid decline in production. It is indigenous to the Amazon Valley (Holliday, 1952) and is now found in Venezuela, Colombia, Peru, Ecuador, Guyana and Surinam on the South American mainland as well as on the islands of Trinidad, Tobago and Grenada. Apart from Surinam, witches' broom has been a major cause of decline in production in Ecuador and Trinidad. The disease does not occur in Bahia,

Brazil, Central America or other West Indian islands and has not spread to West Africa or the Far East.

In Trinidad pod losses of 70 per cent have been noted in uncontrolled areas (Holliday, 1952). Substantial pod losses are experienced in Ecuador, where it is not unusual to find areas with 200 to 300 brooms per tree and a high percentage of pods infected with witches' broom.

Symptoms   The most obvious symptom of the disease, which gives rise to its name, is the characteristic shoots or brooms, caused by the hypertrophic growth of a normal bursting bud. A broom is much thicker than a healthy shoot and carries many short lateral shoots with undeveloped leaves and typically exhibits greatly shortened internodes. Brooms are green at first, but later the host tissue dies and they turn dark brown. Brooms show the leaf arrangement appropriate to the part of the tree attacked and therefore both fan brooms and chupon brooms are seen.

A flower cushion may become infected in which case all flowers and shoots exhibit hypertrophic growth. The colour of infected flowers will initially be normal, but they will have abnormally thickened stalks and sometimes malformed pods will develop. Pods from an infected cushion will die while still quite small.

Infected pods exhibit a variety of symptoms depending on the method of infection and the age of the pod at the time of infection. The most severe infection occurs when a pod is about 1 cm long and causes the pod to develop in a carrot shape and to die when about 15 cm long with the beans a total loss. Pods infected when 2 to 5 cm long become distorted, ripen prematurely and are lost. Larger pods do not become distorted but a dark hard necrotic area develops and again the beans are lost. The least severe infection will occur as pods near maturity when external damage appears as a black speckling, though internally some beans may be lost if harvesting is not prompt. With experience pods infected with witches' broom can be recognised by the extremely hard external necrotic areas, extensive internal necrosis, and the adherence of the beans to the husk (Holliday, 1952).

Epidemiology   The fungus (*Marasmius perniciosus*) is a member of the family which also includes the mushrooms and toadstools. Large numbers of small pink fruiting bodies are formed on the old dead brooms from which vast numbers of spores are liberated. The spores are light and are liberated on dry nights to be carried considerable distances by wind. They are thin-walled and shortlived, the maximum life expectancy being less than forty-eight hours, even under the most favourable conditions. In an environment with relative humidity over 90 per cent germination is very rapid, being almost complete four hours after the spores land on suitable cocoa tissue. Any expanding cocoa tissue can be infected by contact with spores to result in a broom, diseased cushion or infected pod. The fungus does not stimulate a bud to burst, but infects it after bursting. The development of brooms after infection depends on the flushing of the cocoa tree, the largest number of brooms being produced at the time of maximum flush. A vigorously growing tree will tend to produce a large broom. Within the cocoa tree the fungus is localised around the point of infection; it is not systemic and each broom arises from a separate infection.

The broom dies five or six weeks after it has been formed and remains on the tree as a dry broom. There is then a lengthy period of dormancy lasting five to six months, sometimes longer, before the mushrooms are formed. The mushrooms appear in wet weather, very wet days interspersed with dry days seeming to favour mushroom formation.

No intermediate or alternative hosts outside the genus *Theobroma* are known; there are isolated records of fruiting bodies on wood other than cocoa, but these are always in contact with a broom.

Control   There are three approaches to control, the removal and destruction of diseased material, spraying, and the planting of resistant selections.

The removal of all vegetative brooms involves the removal of 15 cm (6 in) of healthy stem; in addition diseased cushions and some bark should be removed, together with any infected pods. This material should be destroyed by burning or burying and none should be left under the trees. This procedure should be carried out at four-month intervals, but in Trinidad the rainfall pattern allows this to be safely reduced to two clearances per annum (Holliday, 1954a). This treatment should effectively reduce the losses of pods but will not eliminate them; it may, however, prove ineffective in old fields of cocoa. This treatment is expensive and especially difficult in tall old trees and such areas should perhaps be replanted. The removal of brooms should ideally be carried out on areas of several hundred hectares at a time. This is often impossible when several holdings are involved, and so success may be limited and short-lived due to rapid re-infection.

Spray trials involving monthly applications of 1 per cent Bordeaux reduced the incidence of both vegetative brooms and infected pods, but some brooms still occurred (Thorold, 1953b). More frequent applications (at one or two week intervals) did not eliminate all brooms and was of little advantage. It was calculated that spraying with copper fungicides is unlikely to be economic unless the pod loss is greater than 15 per cent on trees yielding over 800 kg per hectare (Holliday, 1954b). Spraying with copper is not recommended in any cocoa area, even when severe infections exist.

The ultimate answer to control of witches' broom can only be the universal planting of resistant high-yielding material. The search for trees resistant or immune to witches' broom and the selection and breeding work that followed has been described in Chapter 5. This lengthy programme yielded selections with considerable resistance and high yield and they have been used in new plantings in Trinidad. This material was sent from Trinidad to Ecuador but after good perform-ance initially the same selections became badly infected with witches' broom which indicates some change on the part of the fungus. This shows that some fresh source of resistance must be found and used in a breeding programme.

Cushion gall disease (*Calonectria rigidiuscula*)

Situation and outlook   Cushion gall is a collective term for a number of forms of flower cushion hypertrophy. It seems unlikely that all forms have the same causal organism, but the pathogen for the two most important—greenpoint and flowery galls—has been identified as mat-ing types of *Fusarium rigidiuscula* (Snyder *et al.*, in press). *Calonectria rigidiuscula*, the perfect forms of this fungus, had previously been considered as the causal organism. Brunt and Wharton (1962) studied gall formation and isolated a gall-inducing strain of *C. rigidiuscula* as the pathogen in Ghana. This fungus is widely found in Ghana as a weak parasite, but rarely forms galls; several isolates may in fact be present, some of which are gall inducing and some of which are weakly parasitic and are associated with cocoa dieback. The other forms of gall may be caused by other strains of the same fungus or by different species. Cushion galls in some form are known in most cocoa-growing countries and Hutchins and Siller (1960) proposed names for and described the various types.

Galls are predominantly important in South and Central America but it is difficult to assess the economic importance of cushion gall disease, because incidence varies greatly by area and planting material and the various reports do not always use the same nomenclature.

Severe infections of up to 75 per cent of trees have been reported from parts of Nicaragua and Colombia but these are unusual. In one planting in Costa Rica 13 per cent of the trees were infected but in most other countries in the Caribbean area infected trees are relatively uncommon (Hutchins, 1958). Cushion gall occurs in West Africa and has been known in Ghana since 1923 but infected trees are rare; a survey of 200,000 hectares in 1960 only revealed forty trees with galls (Brunt and Wharton, 1961). In Costa Rica infected trees have been found to produce only half as many pods as healthy trees (Siller, 1961). This no doubt results from infected cushions being sterilised by the disease.

Cushion gall is a widespread disease which has generally been of little economic importance. It appears that an epidemic occurred in certain parts of Central and South America during the years 1955–60.

There have been few reports on this disease later than 1960 and it seems that the epidemic has passed.

Symptoms   Five types of cushion gall have been described (Hutchins and Siller, 1960).

Green-point gall. The swollen flower cushions produce numerous flower initials, but these never develop into normal flowers with stalks and the buds remain green and unopened. These give the appearance of a profusion of 'green-points' on the brown surface of the gall. A gall will be bright green at first, dying after about twelve months and all the tissues then become black and crumble easily. The galls are borne on a short central stalk about 1 cm in diameter and the surface of the gall will be 10 to 15 cm in diameter. The internal tissues are light in colour and soft in texture, branching laterally from the central stalk. The lower surface will probably be flattened against the trunk without adhering to it. These galls typically occur on a flower cushion, though exceptionally they develop elsewhere.

Flowery gall. With this type of gall cushions may bear hundreds of closely packed flowers, in a series of flushes through the flowering season. The flowers develop fully. Soria (1960) showed by hand pollination that the pollen is viable and the ovules are capable of producing normal seeds, so the lack of fruit setting is probably due to failure of pollination. The internal structure of the flowery gall is similar to the green-point gall, with a stem connected directly with the wood of the supporting branch. This condition can be confused with the usual phenomenon of profuse flowering, probably associated with self-incompatibility or sterility. This distinction is especially difficult when the gall is in the early stages of development. Marked differences in varietal susceptibility have been reported (Hutchins *et al.,* 1959).

Knob gall. These are hard, woody, smooth-surfaced swellings up to 2 cm or more in diameter, which may occur in the flower cushions, but bear no flowers. They are widely distributed but relatively unimportant. An affected tree seldom shows more than 10 to 15 of these galls.

Disc gall. Disc gall, formerly called 'hard flat gall' was first noted in Ecuador and Brazil in 1958 and has been found in Guatemala. These galls are woody, very hard and firmly united with the wood of the supporting tree across the whole diameter of the gall.

Fan gall. This name was given to cushion galls, usually of the flowery type, which occasionally develop stemlike growths of up to a few inches in length in the shape of a fan.

Control   Until the causal organisms of these five types of galls have been clearly isolated sound control methods cannot be formulated. Hardy (1960) considered that if only a few gall-bearing trees are found they should be removed and destroyed at once and recommends regular inspection in susceptible areas. Hutchins (1958) suggests that cushion gall diseases could be introduced with planting material, and recommends that propagating material should only be taken from healthy trees devoid of galls.

Mealy pod disease (*Trachysphaera fructigena*)

This fungus belongs to a genus closely allied to *Phytophthora* and the symptoms produced in infected pods are very similar to those of black pod disease. The edges of the lesions are usually diffuse whereas black pod lesions are clearcut, but this is not a reliable means of distinguishing the two diseases.

In Ghana losses to mealy pod are generally small. Legg (1970) estimated losses at 3 per cent between October 1968 and January 1969; Dakwa (1972) found evidence of higher losses between July and November. Losses up to nearly 8 per cent have been recorded in Nigeria (Weststeijn, 1966).

The fungus is a wound parasite and there are conflicting opinions as to whether it can attack undamaged pods; Dakwa (1972) suggests that gross damage is required for entry of the fungus. It is doubtful whether losses are high enough to warrant any special control measures. Regular and frequent harvesting, coupled with the control of rodents, will help to prevent losses from mealy pod.

Botryodiplodia pod rot (*Botryodiplodia theobromae*)

This disease is also called diplodia pod rot and brown pod rot; the latter name can cause confusion with the much more important black pod disease as the French and Portuguese names for black pod disease are literally translated as brown pod rot. Furthermore both diseases start as brown lesions which turn black.

This fungus is ubiquitous and is a weak wound parasite causing minor losses of pods. The lesions produce masses of black spores so that diseased pods become covered with black sooty powder in contrast to the white mycelium of black pod. The disease only occurs on wounded pods or pods which are under stress. It tends therefore to be found more frequently in the dry season and has been found in a large proportion of diseased pods in the dry season in Ghana and Papua–New Guinea (Legg, 1970; Thrower, 1960). Losses are not great enough to justify control measures.

Other pod diseases

Several other fungi have been reported as attacking cocoa pods but all are localised and of minor incidence.

In Malaysia *Phytophthora heveae* has been reported from isolated attacks (Turner, 1968); on the other hand *P. megasperma* causes severe losses in eastern Venezuela (de Reyes *et al.,* in press). In Costa Rica *Fusarium roseum* has been found on cherelles and may remain dormant in them causing internal rot when the pod matures (Waite and Salazar, 1966). Other *Fusarium* species attack pods in West Africa causing minor losses.

Diseases affecting the trunk and canopy

Ceratostomella wilt (*Ceratocystis fimbriata*)

Situation and outlook   This disease occurs as a wound parasite associated with damage by implements or beetle borers. The wilting and death of branches and of whole trees in many countries has been known under the name of Ceratostomella disease, wilt disease or the *Xyleborus-Ceratocystis* complex. The association between the disease and Xyleborus beetles has been well documented by Iton (1959, 1961) working in Trinidad, where wilt disease is important. Infection is also associated with cutlass and pruning wounds and it is probable that such wounds are important means of primary infection of a tree.

The first outbreak of wilt disease probably occurred in Ecuador in 1918, but was considered to be unimportant because the prevailing variety at that time was Nacional, which is resistant to this disease. It is however possible that the severe 'blast' which devastated the Trinidad cocoa industry in 1727 may have been due to this fungus. In 1951 according to Desrosiers (1957) a new and more virulent form of this pathogen was detected at the Pichilingue Research Station in Ecuador. It was noted that those trees with predominantly Criollo characteristics were much more seriously affected than Trinitario and Forastero varieties. In 1956 small groups of trees on estates in widely separated localities in Trinidad were dying and this prompted the most recent investigations into the nature of wilt disease.

The disease occurs in Colombia, Ecuador and Venezuela, in which countries it has caused serious losses, and has also been recorded in Trinidad and Costa Rica, Guatemala, Hawaii, Dominican Republic and the Philippines (Entwistle, 1972). *Ceratocystis fimbriata* occurs in West Africa on hosts other than cocoa, but wilt disease as we know it has not been reported from West Africa.

It is likely that the disease has been present in many cocoa-producing countries for many years, but has only become severe when some condition of stress predisposes the trees to attack. It is therefore very difficult to assess the direct economic effect of this pathogen, but Saunders (1965) considered that millions of trees have been killed by it. *Ceratocystis fimbriata* is so widely distributed that wilt disease is a potential hazard wherever cocoa is grown.

Symptoms   The external symptoms are wilting of the whole or part of the tree followed by rapid death of the affected part. In the earliest stage the mature leaves change from the normal horizontal position to a pendulous one, similar to that of young leaves which have just completed flushing. These wilted leaves become brown and rolled longitudinally and characteristically remain attached to the dead branch in a shrivelled condition for several weeks. This condition is always associated with borings of the trunk or branches of the tree. These holes, made by *Xyleborus* spp., are characteristic and usually 1 mm in diameter, penetrating the wood at right angles to the axis and typically exhibiting a small amount of wood dust immediately around the hole (Iton, 1959). The internal wood tissue surrounding the wound will be discoloured brown-red or purplish with the colour decreasing towards the healthy areas and it may be that this variation in colour is due to the presence of other fungi as secondary invaders (Iton and Conway, 1961). The extent of the wilting is related to the siting of the boreholes.

Considerable time elapses between infection and the appearance of visible signs of attack. Spence and Moll (1958) examined externally healthy trees and found extensive damage to bark and wood within the tissues before the wilting of trees was visible. Not all borer damage to cocoa trees is associated with Ceratocystis wilt even in areas of heavy infection. Spence and Moll suggest that the outbreak of wilt in Trinidad in 1956 reflected the stress that cocoa trees had suffered through rainfall deficiency in the previous months. It seems highly likely that ecological factors such as drought or flooding can indeed weaken trees and thus initiate a cycle of wilt infection.

Epidemiology   The fungus produces its spores largely, if not entirely, within the tree, especially in the galleries of the *Xyleborus* beetles. They are exuded from the tree with the frass or wood dust and are spread by the wind and by the *Xyleborus* beetles themselves. It is possible that other insects—mites, nematodes and springtails—are also involved (Iton, 1966). At River Estate in Trinidad there was clear evidence of spread of disease in the direction of the prevailing wind, but jump spread also occurs which might result from infection from other host plants.

Spores can infect cocoa trees through cuts made by a cutlass or machete, hence the name *mal de machete* in Latin America. In the association of the fungus with *Xyleborus* beetles, there has been difficulty in determining whether the fungus or the beetle attacks first. In Trinidad the evidence points to infection by the fungus preceding attack by the beetle (Iton and Conway, 1961), and it has been shown that the beetles show a preference for newly infected trees rather than healthy ones (Iton, 1959). On the other hand in Ecuador and Costa Rica there is clear evidence that beetle attack precedes infection.

Control   Neither chemical control of beetle or fungus nor destruction of infected material have proved successful as control methods. The removal and burning of all infected branches and dead trees has been recommended but this may in fact disturb beetles and infected debris and spread the spores to healthy trees. The most useful technique is to prevent the spread of fungus by minimal damage at pruning and harvesting. Sterilising the cutlass after each tree by incorporating a fungicide in the scabbard or using protective fungicide paint on a large exposed surface may also be useful in control.

Many workers have indicated that Criollo types are more susceptible than Forastero selections, but there is also considerable variation within Criollo populations. Resistance has been found in IMC 67 and other selections and this has been used successfully in the breeding programme of the Cocoa Board in Trinidad (Freeman, in press).

Dieback

Situation and outlook Dieback is a general condition of the cocoa tree due to one or more of a number of different causes, some of them physiological, others pathological. Before the pathogens were isolated both vascular streak dieback (*Oncobasidium theobromae*) in the Far East and *Verticillium* wilt (*Verticillium dahliae*) in Uganda were classified as dieback. It is possible that future work will isolate other specific pathogens from the broad classification of dieback. Vascular streak dieback and *Verticillium* wilt are considered later, while this section broadly considers the general condition.

The term dieback generally refers to the condition of progressive desiccation of the branches from the tip inwards to the jorquette. The severity of the damage can vary from a few twigs to complete tree mortality and the word dieback has become associated with any tree mortality for which no obvious cause can be seen. The subject has been reviewed in detail by Turner (1967), who states that the condition of dieback has been recorded in nearly every major producing country. It is obviously difficult, in view of the background to this condition, to make an economic assessment of the importance of dieback except to say that it has been responsible for the death of many trees in various parts of the world and for considerable losses of production in addition to rendering some areas uneconomic for cocoa cultivation.

Symptoms The disease is characterised by death of tissues beginning at the tips of the finer branches of both young and old trees and gradually extending back towards the jorquette. The first external sign is usually the twig withering prior to the leaves turning brown, and the progress of the disease is shown by this continuous drying of the leaves beginning with the youngest. The leaves are usually shed and the replacement leaves are also shed, until only a bare tree remains. The speed of this process is variable, but the whole tree often appears unhealthy from an early stage.

Causes of dieback   The disorder may be caused by any factor which influences yield. These include environmental, physiological and nutritional disturbances, fungal invasion and pest attack as well as interactions between any of these components.

Dieback symptoms during drought conditions have been noted in many countries. It is suggested that dieback may be associated with an annual rainfall of less than 1,400 mm (55 in) per annum, though it is more likely that rainfall distribution is the significant factor. Dieback symptoms have been induced in experimental plants grown under water stress (Turner, 1967). The water relations of an individual plant and its vigour of growth will depend on the nutrient status of the soil and its acidity, salinity, structure and texture. All these factors are interrelated and acting together will influence the incidence of dieback. Trees which are suffering from water stress will succumb more easily to attack by fungi and pests. In addition waterlogging may cause dieback symptoms.

Excessive insolation due to lack of shade has been suggested as a cause of dieback. Shade reduction will result in higher rates of photosynthesis, which will give more vigorous growth and higher yields, where there is adequate soil moisture and nutrients, but where these factors are limiting the trees will be weakened and suffer from dieback. There will obviously be differences between types of planting material in their ability to withstand prolonged insolation or dry season desiccation. Dieback can also result from the effect of strong or drying winds. The harmattan wind in West Africa causes a severe drop in humidity which can lead to defoliation and dieback. In the West Indies the trade winds can bring about similar effects.

Attacks by insects have frequently resulted in the development of severe symptoms of dieback. Insect attack may affect the tree in a number of ways including physical damage, production of toxins, or the physical damage can offer entry to fungi. One or all of these factors may result in the death of the tree or main branches. There is often an improvement of the dieback situation following successful chemical control of pests (Donald, 1957; Kay, 1961).

The insects most frequently associated with dieback are capsids which can cause severe damage by the action of toxins on the tissues around the punctures but this damage may be aggravated by the invasion of the lesion by the fungus *Calonectria rigidiuscula*. Other insects associated with dieback include *Earias biplaga*, which attacks young shoots in unshaded cocoa in West Africa, and *Selenothrips rubrocinctus*, the red banded thrips, which is widespread and attacks unthrifty cocoa.

Toxicity and deficiency of mineral elements can also cause dieback symptoms in addition to a reduction of vigour and specific metabolic disorders. The symptoms of nutritional disorders are mentioned in Chapter 9.

Very many fungi have been isolated from tissues affected by dieback and over eighty are listed by Turner (1967), but not all are pathogens. It is suggested that when a tree becomes weakened by nutritional or environmental factors the tips of finer branches die, thereby creating a suitable substrate for a wide range of saprophytic fungi. At the same time the reduced vigour of the tree will lower its resistance and predispose it to attack by weak parasites. *Botryodiplodia theobromae* is probably a secondary invader though many authors suggest it to be a primary pathogen. Some isolates of *Calonectria rigidiuscula* are gall inducing, while others act as a weak parasite (Turner, 1967).

Control The first step in attempting to control dieback is to ascertain the cause. In some cases the attack may be due to exceptional climatic conditions or some other temporary reason and the attack can be expected to disappear. In other cases a more permanent cause may be identified and some control measures should be considered. Where the attack is due to water stress some means of alleviating it must be sought; this might involve irrigation, the planting of shade trees or a wind break, closer planting of the cocoa trees or possibly the removal of competing weeds. Adequate nutrition of the trees should be ensured which may involve application of fertilisers or micronutrients where deficiencies are implicated. Where dieback is due to insect damage, the use of suitable pesticides should lead to an improvement. It is obvious that there are many ways in which dieback has to be countered as there are many causes of the dieback condition. In some cases the control measures will prove fully effective, in others only partially effective and occasionally the problem will require detailed research to establish the cause and suggest a control.

Vascular streak dieback (*Oncobasidium theobromae*)

A destructive form of dieback has recently been found to be caused by *Oncobasidium theobromae*, a previously undescribed species of fungus, which occurs in Papua–New Guinea and Malaysia (Keane *et al.*, 1971). It has been called vascular streak dieback because streaking of the woody or vascular tissue is a characteristic symptom.

The incidence of this disease became serious in Papua–New Guinea about 1959, particularly in high rainfall areas, and it may even be that this pathogen renders cocoa cultivation uneconomic in areas of rainfall over 3,000 mm (120 in). A dieback which appeared in West Malaysia in 1957 and caused a severe setback to cocoa-growing, has proved to be vascular streak dieback (Keane and Turner, 1972). It is widespread in West Malaysia but only severe in its effects on the east coast where the rainfall is heavy. The severe effects are found on plantings of Amelonado which has proved to be far more susceptible than Amazon material.

Symptoms   The first indication of the disease is a characteristic yellow-ing of one or two leaves on the second or third flush behind the growing tip. This develops into small sharply defined green spots scattered over a yellow background; diseased leaves fall within a few days of turning yellow. The other leaves on the shoot soon show similar symptoms. As the leaves fall, short lateral shoots grow from the leaf axils. Un-hardened leaves in the young flush of a diseased shoot may show an oakleaf pattern due to tissue death between the lateral veins.

If the dry surface layer is removed from a leaf scar after the fall of a diseased leaf the black vascular bundles will be seen. When an infected shoot is split lengthwise there is always a characteristic brown streaking of the woody tissue extending well beyond the region of yellowed leaves. The disease eventually kills the shoot and if unchecked will extend back along the main branch resulting in progressive death of the tree. Leaf yellowing symptoms will appear in the field three months after the spores alight on the tree (Keane *et al.*, 1971).

Epidemiology   The disease is spread by the spores produced on dis-eased branches which are released only at night under specific climatic conditions and are dispersed by wind. A spore which is deposited on a young flush can produce a new infection. The distance over which spores can be carried will be limited by the lack of winds at night and possibly by the high humidity which causes condensation on the spores to increase their weight. The spores are killed by bright sunlight and fruit bodies probably only shed spores for a few nights. These factors all limit spore dispersal.

Control   The regular cutting and disposal of diseased branches will remove the source of infection and regular pruning of chupons on the trunk will deny the fungus access through these unhardened leaves. Cocoa nurseries should not be sited near diseased trees.

The identification of the pathogen has enabled selection of resistant material to be developed and selection programmes have been started in Papua–New Guinea as well as Malaysia. It has already been stated that Amelonado types are highly susceptible to this disease; Amazon varieties show symptoms but continue to yield well in areas where the disease is prevalent. This forms the basis for selection of resistant material in West Malaysia. Greater knowledge of the fungus and its epidemiology may lead to chemical control, but the selection of resistant material appears to be the more promising approach.

Sudden death disease or *Verticillium* wilt (*Verticillium dahliae*)

Situation and outlook   A dieback or sudden death disease has occurred on cocoa in Uganda for many years and may well have been a cause of the failure of cocoa to become a significant crop in that country

(Leakey, 1965). With the revival of interest in cocoa following the introduction of Amazon planting material in 1956 the disease reappeared and was studied. It was now attributed to *Verticillium dahliae* and is referred to as *Verticillium* wilt (Emechebe and Leakey, 1968). Incidence of the disease has risen as high as 30 per cent in one area and it is the major disease affecting cocoa in Uganda (Emechebe *et al.*, 1971).

Symptoms   The commonest external symptoms in the field are the sudden wilting and subsequent death of the leaves. Initially the leaves droop without loss of turgor but retain their colour. The leaves start to desiccate beginning from the tips and leaf margins, while the margins start to roll inwards. The leaves eventually become brittle but will remain attached to the stem for over five weeks. Gradually the finer branches break off and the leaves fall, so that the affected shoot becomes completely devoid of leaves and fine branches (Emechebe *et al.*, 1971). Severe attacks following very dry or especially wet weather can kill the trees within a week.

The most distinct internal symptom is the discoloration of the petiole, pedicel, stem and roots. The colour will initially be light brown, but before the appearance of any external symptoms, most of the discoloured areas will be dark brown. This colour change is not uniform but occurs in distinct streaks and is associated with a blockage in the vascular tissues of the plant (Emechebe *et al.*, 1971). Such an obstruction will reduce the yield potential even if death does not follow.

Control   Well shaded cocoa is not as severely affected by this disease as unshaded cocoa (Trocme, 1972). This suggests that adequate shade may be a useful method of control in Uganda. Chemical control has not been tested, but treatment with anything other than a systemic fungicide is unlikely to be useful as the action of the fungus is internal. Selections resistant to the disease may well be found and might provide a long-term solution.

Pink disease

Situation and outlook   This disease is insignificant in all the major cocoa-producing countries, but causes damage in some isolated areas and is severe in parts of Malaysia. It is caused by *Corticium salmonicolor*, a fungus which is of considerable importance on rubber in the Far East. This fungus has a very wide host range, entirely amongst Dicotyledons, and has been recorded on at least 104 different genera including rubber and a number of other crops of economic importance (Briton-Jones, 1934). It has been recorded on cocoa in Ghana, Nigeria, Cameroon, Malaysia, Papua–New Guinea, Western Samoa and Trinidad. In New Guinea and Malaysia pink disease is associated with the use of cover crops such as pigeon peas (*Cajanus cajan*), *Crotalaria* and *Tephrosia* spp.

The economic significance of this disease on cocoa is hard to assess, but in parts of Sabah frequent spraying combined with removal and disposal of externally damaged tissues is considered necessary and in one instance 35 per cent of trees under four years have been attacked. Both fan branches and the main stem below the jorquette are prone to attack by the fungus. Severe infection of the main stem in trees under two years of age frequently results in death of the tree, though a vigorous tree may recover through rapid development of healthy chupons from lateral buds below the infected area.

Symptoms   Unfortunately the disease is not readily detected until it has penetrated deep into the bark and the first visible sign is usually the appearance on the bark of the salmon pink incrustation of fruiting bodies from which the disease derives its name. Thus detection is only possible after the most vigorous phase of infection, when the many fine silky white mycelia have already spread over the surface and into the cortex of the bark. The pink incrustation consists of millions of spores, and their release coincides with the death of the host tissue as shown by cracking of the bark. The penetration of the fungus to the cortical tissue disrupts the physiological processes of the tree and quickly leads to defoliation and death of the distal parts of the branch. Prolonged insolation will bleach the pink colour of the fruiting bodies to a greyish white colour.

Epidemiology   The spores of the fungus are windborne and spread quickly, active sporulation occurring during the hours of darkness and favoured by conditions of high humidity (Edgar, 1958). The bark of the trees on which spores settle must be moist enough for germination of these spores and much can be done by cultural measures to lower the humidity within the canopy to reduce germination. It has been noted that infections of this pathogen are more serious in areas of open canopy. It may be that the trees are already weakened by excessive insolation and the attack of pink disease is thus more marked. It is also probable that dispersal of spores will be more efficient in areas of low shade as a greater expanse of woody tissue will be exposed to infection.

Control   For severe infections regular spraying may be necessary. Application of 1 per cent Bordeaux has been shown to be effective for control in Trinidad, but there is no experimental evidence for other areas (Thorold, 1953b). It should be stressed that good surface drainage combined with adequate shade, as opposed to overshaded conditions, can markedly reduce the humidity within the canopy and thus the opportunity for germination, by shortening the periods of high humidity following heavy rain. There is no experimental evidence to suggest the ideal shade regime for areas of high infection of pink disease. This pathogen can also infect a variety of jungle trees and cover crops and it is thus desirable that no susceptible species should be used as shade or ground cover.

Thread blights (*Marasmius* spp.)

Situation and outlook   Thread blights occur in most cocoa-growing countries. There are two types—the white thread blight caused by *Marasmius scandens* and horsehair blight caused by *M. equicrinus*. The damage caused by thread blights is considered to be unimportant and their incidence and effect on yield have been little studied. Random samples of trees in Ghana showed an incidence varying from 6 to 48 per cent infected trees (Leston, 1970).

Symptoms   White thread blight kills leaves and a network of mycelial threads spreads over leaves, petioles and branches. The dehisced leaves remain hanging to the branch by strands of mycelia.

Horsehair blight forms a tangle of black fungal threads through the canopy. The leaves are not killed, but after natural dehiscence are held to the tree by the fungal threads and thus tend to smother new growth.

Control   Damage from these diseases can be reduced by removal of the dead material and pruning of affected parts. Further pruning may be necessary to reduce the humidity in the canopy and in severe cases it may be necessary to use a copper spray and at the same time remove the infected wood. It is doubtful, however, whether spraying would be worth while as there is no indication that control of thread blights leads to an increase in yield.

Root diseases

Situation and outlook   This group of diseases is rarely responsible for tree losses greater than 1 to 2 per cent per annum in any planting. In rubber plantations root diseases can obliterate large areas completely, but in cocoa plantings the losses tend to be of isolated trees randomly distributed throughout the area.

Root diseases are caused by a number of pathogens, whose relative importance varies from country to country. The original source of all root diseases of cocoa will be the forest cleared prior to planting and a period of cultivation of non-woody crops may reduce a severe level of inoculum to an insignificant level. The first symptom noted for all root diseases is the rapid wilting of leaves on a single tree followed quickly by death of the tree. In the absence of other aerial symptoms, such as borers, an attack by root disease is the most likely cause of such symptoms.

Control   There is little prospect of chemical control of these pathogens; in general they are underground and frequently infection can proceed along the root internally and thus avoid any fungicidal dressing. An infected tree generally dies before its neighbour becomes infected and so prompt removal of a dead infected tree can check further spread.

Brown root disease (*Phellinus noxius*)

This pathogen is probably more widely known under its old name *Fomes noxius*. This pathogen attacks cocoa in a number of countries including Ghana, Nigeria, Sri Lanka, Malaysia, Papua–New Guinea and Samoa, and attacks a wide range of other species including cola, rubber and tea.

Symptoms   An obvious feature of roots infected with this pathogen is the hard brittle encrustation of soil round the root, held by exudate from the profuse brown rhizomorphs. These rhizomorphs soon develop into a complete fungal skin and eventually turn black. At the earliest stages of infection before tree death the mycelium will appear golden and the infected wood is at first brown. At an advanced stage of infection a honeycomb of lines can be seen below the bark and the wood is then friable. The fruiting body is rarely seen but its morphology is well described by Briton-Jones (1934).

In Papua–New Guinea this fungus is the most important root disease and there is evidence that in that area it has two distinct sets of symptoms (Thrower, 1965). The second manifestation of the fungus consists of a crust of mycelium on the lower trunk up to a height of one metre, which appears unconnected with mycelium in the roots, thus suggesting it has arisen from a separate infection.

Epidemiology   The disease spreads very slowly because infection only occurs by direct contact with infected roots, which are often part of decaying forest trees. The aerial symptoms found in Papua–New Guinea could result from knife wounds on the trunk, the pathogen acting as a wound parasite.

White root disease (*Rigidoporus lignosus*)

The causal organism is another fungus, which has a variety of names. The most recent is *Rigidoporus lignosus*, but it is better known under the name *Fomes lignosus*. Its greatest economic importance is as a root parasite on rubber in the Far East, but it has a wide range of other host plants. It must still be classed as a minor disease of cocoa even in the Far East, but it could potentially become serious on estates where rubber and cocoa are mixed.

Symptoms   The wilting of leaves and dying of branches may be sudden and can be completed in a few days or it may take several weeks. The diseased roots will be covered by white rhizomorphs which are strongly fixed to the roots, and age to an orange-red colour. Fruiting bodies are frequently seen after the death of the tree at the collar in the form of large brackets with orange-yellow upper surface and orange, red or brownish lower surface, while in cross section the bracket offers internally an upper white layer and lower red-brown layer. Some harmless saprophytes also produce white rhizomorphs, but these are usually loosely attached to the root (Edgar, 1958).

Epidemiology   The spread of the disease is by direct root contact with an infected tree although there is a possibility that windblown spores can infect wood recently exposed after injury (John, 1964).

Collar crack (*Armellaria mellea*)

This fungus attacks many species of trees in both tropical and temperate regions and is commonly known as the Honey Agaric. It occurs in Ghana, Nigeria, Cameroon and Togo, attacking cocoa trees of any age; the attacks are almost always fatal. The fungus flourishes in damp conditions.

Symptoms   The characteristic fruiting bodies are in general clustered at the base of the tree and are of a light brown colour at first turning yellow and finally black. The mycelium from the fruiting bodies invades the rest of the tree, especially the medullary rays. This causes the rays to thicken and split the wood to give a 'collar crack' in the stem tissue. Such a crack can be one or two metres in length. Progress of the attack is rapid. The whole tree eventually falls, but wilting may not occur until the tree has fallen (Wharton, 1962).

Epidemiology   Black rhizomorphs are usually associated with this fungus, but have not been noted on cocoa. Infection spreads from tree to tree by contact with adjacent root systems; damp conditions or waterlogging of roots can probably predispose a cocoa tree to infection from this pathogen (Dade, 1927; Weststeijn, 1967).

There is a wide range of host plants for the pathogen in Ghana (Dade, 1927). Hardwood trees are the most dangerous species as sources of inoculum; they are frequently felled before planting cocoa and their roots survive for many years. Softwood types in a humid environment are quickly invaded by saprophytes which prevents colonisation by the pathogen.

Black root disease (*Rosellinia pepo*)

Three species of the genus *Rosellinia* have been found to attack cocoa, but only *R. pepo* is important and is the commonest root disease of cocoa in the West Indies (Briton-Jones, 1934). It attacks a number of other crops and tends to occur in patches which spread slowly, though the attacked tree may die suddenly. The pathogen is a normal inhabitant of forest soils and infection usually arises from the stump of a forest tree.

Symptoms An infected root will exhibit a covering of a smoky-grey mycelium turning purplish black, with white mycelial fans beneath the surface of the root bark. An infected tree will wilt, its leaves will die and the death of the tree will eventually follow. Sections through old infected wood show the presence of the fungus as thin black lines in the vascular tissue (Briton-Jones, 1934).

Mistletoes

Situation and outlook At least six species of mistletoe have been found on cocoa in West Africa. In Ghana about 70 per cent of all individuals are *Tapinanthus bangwensis*, which is characterised by red flowers, red berries and a globular haustorium; a further 20 per cent are *Phragmanthera incana*, which has yellow flowers, blue fruit and an elongated haustorium; the remaining 10 per cent comprises four other species (Room, 1972). Mistletoes have not been recorded from cocoa-producing countries outside West Africa other than Costa Rica, where they are considered to be a major cause of damage (Kuijt, 1964).

The biology and ecology of *T. bangwensis* has been investigated in some detail and there is evidence suggesting that mistletoe infestation of cocoa in Ghana and Nigeria is increasing (Room, 1972). This may be accounted for by the general reduction of shade in these countries but it is possible that mistletoes have been slowly adapting to a new host.

Effects of T. bangwensis on cocoa Mistletoes have no normal roots and therefore rely entirely on their hosts for water and mineral salts, they also act as sinks for the products of photosynthesis manufactured by cocoa (Room, 1972). In addition they damage the canopy by killing cocoa branches distal to themselves.

The mistletoe fauna is extremely complex. All mistletoes are attacked by wood-borers, whose galleries are then occupied by a mealybug specific to mistletoes (*Cataenococcus loranthi*). This mealybug is not implicated in the transmission of the cocoa swollen shoot virus but is tended by a species of ant which also tends mealybugs carrying virus. Therefore mistletoes indirectly encourage the spread of swollen shoot.

Life cycle   *T. bangwensis* flowers twice a year, once in March/April and again in July/August, when all plants bear masses of bright red flowers which are pollinated by sunbirds and bees. The bright red berries take one month to ripen and are eaten by various species of birds. The seeds pass through the gut of the bird very quickly, retaining an adhesive coating, which ensures that the birds have to wipe them on to a suitable twig. The seeds will only germinate in unshaded conditions and can only penetrate the relatively thin bark of young branches. Young mistletoes can flower about nine months after germination and the maximum life span is about eighteen years.

Control   The maintenance of top shade to prevent mistletoe germination is a useful long-term measure, but for immediate results manual cutting out is still the best method. A long-handled pruning knife is ideal for this work and branches should be severed some 10 mm proximal to the mistletoe to remove completely all traces of the parasite. In Ghana the best time of year to carry out this work is during the August flowering peak, when the plants are easily seen in the canopy and farmers are not involved in harvesting (Room, 1972). *T. bangwensis* has a very wide range of alternate hosts and outside sources of infestation are numerous. Nevertheless the reproduction rate of this mistletoe is sufficiently low to make treatment every other year effective in controlling these parasites.

References

Introduction:
Cramer, H. H. (1967) 'Plant protection and world crop production', *Pflanzenschutz — Nachrichter 'Bayer'*, **20** (1).
Hale, S. L. (1953) 'World production and consumption 1951 to 1953', *Rep. Cocoa Conf. London 1953*, pp. 3–11.
Padwick, G. W. (1959) 'Plant diseases in the colonies', *Outlook on Agriculture*, **2** (3), 122–6.

Viruses:
Bigger, M. (1972) 'Recent work on the mealybug vectors of cocoa swollen shoot disease in Ghana', *PANS*, **18** (1), 61–70.
Crowdy, S. H. and Posnette, A. F. (1947) 'Virus diseases of cacao in West Africa, II. Cross-immunity experiments with viruses 1A, 1B and 1C', *Ann. appl. Biol.*, **34** (3), 402–11.
Dale, W. T. (1962) 'Virus diseases', in J. B. Wills, ed., *Agriculture and Land Use in Ghana*, Oxford University Press, pp. 286–316.
Entwistle, P. F., Johnson, C. G. and Dunn, E. (1959) 'New pests of cocoa (*Theobroma cacao* L.) in Ghana following applications of insecticides', *Nature, Lond.*, **182**, 1463–4.
Kenten, R. H. and Legg, J. T. (1971) 'Varietal resistance of cocoa to swollen shoot disease in West Africa', *FAO Plant Prot. Bull.*, **19**, 1–11.

Legg, J. T. (1972) 'Measures to control spread of cocoa swollen shoot in Ghana', *PANS*, **18** (1), 57–60.

Owusu, G. K. (1971) 'Cocoa necrosis virus in Ghana', *Trop. Agric.*, **48**, 133–9.

Posnette, A. F. (1940) 'Transmission of swollen-shoot', *Trop. Agric.*, **17**, 98.

Posnette, A. F. (1947) 'Virus diseases of cacao in West Africa. I. Cacao viruses 1A, 1B, 1C and 1D', *Ann. appl. Biol.*, **34** (3), 388–402.

Steven, W. F. (1936) 'A new disease of cocoa in the Gold Coast', *Gold Coast Farmer*, **5**, 122 and 144.

Thresh, J. M. (1958) 'Virus research. Nigerian isolates of cacao viruses', *Rep. W. Afr. Cocoa Res. Inst. 1956–57*, pp. 71–3.

Thresh, J. M. and Tinsley, T. W. (1958) 'Virus diseases of cacao: the world situation', *Septima Conf. Inter-Amer. de Cacao, Palmira, Colombia 1958*, pp. 201–17.

Thresh, J. M. (1960) 'Capsids as a factor influencing the effect of swollen-shoot disease on cacao in Nigeria', *Emp. J. exp. Agric.*, **28** (3), 193–200.

Black Pod:

Atanda, O. A. (1973) 'Breeding cocoa for different ecological needs in Nigeria', *Cocoa Growers' Bull.*, **20**, 17–24.

Blencowe, J. W. and Wharton, A. L. (1961) 'Black pod disease in Ghana: incidence of the disease in relation to levels of productivity', *Rep. 6th Commonwealth Mycol. Conf., Kew 1960*, pp. 139–47.

Braudeau, J. and Muller, R. A. (1971) 'Des possibilités d'emploi des composés organo-stanniques contre la pourriture brune des cabosses du cacaoyer due au *Phytophthora palmivora* (Butl.) Butl. au Cameroun oriental', *Café Cacao Thé*, **15**, 211–20.

Briton-Jones, H. R. (1934) *The Diseases and Curing of Cocoa*, Macmillan, London.

Chant, S. R. (1957) 'A dieback of cacao seedlings in Nigeria caused by a species of *Phytophthora*', *Nature, Lond.*, **180**, 1494–5.

Dade, H. A. (1927) 'Economic significance of cacao pod disease and factors determining their incidence and control', *Bull. Dept. Agric. Gold Coast*, **6**.

Dade, H. A. (1928) 'The relation between diseased cushions and the seasonal outbreak of "Black Pod" disease of cacao', *Gold Coast Agric. Year Book 1927*, Paper 13.

Evans, H. C. (1973) 'New developments in black pod epidemiology', *Cocoa Growers' Bull.*, **20**, 10–16.

Filani, G. A. (1972) 'Studies on the chemical control of *Phytophthora* pod rot', *Ann. Rep. Cocoa Res. Inst. Nigeria 1970–71*, pp. 125–6.

Gorenz, A. M. (1972) 'Field spraying for control of *Phytophthora* pod rot—1970', *Ann. Rep. Cocoa Res. Inst. Nigeria 1970–71*, pp. 115–18.

Gregory, P. H. (1969) *Black Pod Disease Project Report*, Cocoa, Chocolate and Confectionery Alliance, London, 1969.

Gregory, P. H. (1972) 'Cocoa: the importance of black pod disease', *SPAN*, **15** (1), 30–1.

Hislop, E. C. (1963) 'Studies on the chemical control of *Phytophthora palmivora* on *Theobroma cacao* in Nigeria, V. Comparisons of three spraying machines for applying fungicides', *Ann. appl. Biol.*, **52**, 481–92.

Lellis, W. T. (1952) 'Temperaturas como factor limitante da "Podridao parda" dos frutos do cacaueiro', *Bol. Tec. Inst. Cacau, Bahia*.

Manço, G. R. (1966) '*Phytophthora palmivora* in flower cushions, old infected pods and leaves of cocoa plants', *Turrialba*, **16**, 148–55.

Miranda, S. and Cruz, H. M. da (1953) 'Fighting brown pod rot disease in Bahia', *Rep. Cocoa Conf. London 1953*, pp. 120–2.

Muller, R. A. and Njomou, S. E. (1970) 'Contribution à la mise au point de la lutte chimique contre la pourriture brune des cabosses du cacaoyer (*Phytophthora palmivora* (Butl.) Butl.) au Cameroun', *Café Cacao Thé*, **14**, 209–20.

Newhall, A. G. (1966) 'When does it pay to spray cocoa?', *Cacao*, **11** (1), 10–12.

Okaisabor, E. K. (1972) 'Control of *Phytophthora* pod rot disease by ground spraying', *Ann. Rep. Cocoa Res. Inst., Nigeria, 1970–71*, pp. 113–15.

Okaisabor, E. K. (In press) 'Ambient and on-tree reservoirs of *Phytophthora palmivora* (Butl.) Butl. in Nigeria', *Proc. 4th Internat. Cocoa Res. Conf.,* Trinidad, 1972.

Orellena, R. G. and Som, R. K. (1957) 'Correlation between low temperatures and incidence of *Phytophthora* pod rot of cocoa in Ceylon', *FAO Plant Prot. Bull.,* **6**, 6–8.

Padwick, G. W. (1959) 'Plant diseases in the colonies', *Outlook on Agriculture,* **2** (3), 122–6.

Swarbrick, J. T. (1965) 'Estate cocoa in Fernando Po', *Cocoa Growers' Bull.,* **4**, 14–19.

Tarjot, M. (1967) 'Etude de la pourriture des cabosses due au *Phytophthora palmivora* en Côte d'Ivoire', *Café Cacao Thé,* **11**, 321–30.

Tarjot, M. (1971) 'Nouvelle contribution a l'étude de la pourriture des cabosses du cacaoyer due au *Phytophthora palmivora* (Butl.) Butl. en Côte d'Ivoire', *Café Cacao Thé,* **15**, 31–48.

Thorold, C. A. (1953a) 'The control of black pod disease of cocoa in the Western Region of Nigeria', *Rep. Cocoa Conf. London 1953,* pp. 108–15.

Thorold, C. A. (1959) 'Methods of controlling black pod disease (caused by *Phytophthora palmivora*) of *Theobroma cacao* in Nigeria', *Ann. appl. Biol.,* **47** (4), 708–15.

Thorold, C. A. (1967) 'Black pod disease of *Theobroma cacao*', *Rev. appl. Mycol.,* **46**, 225–37.

Tollenaar, D. (1958) '*Phytophthora palmivora* of cocoa and its control', *Neth. J. agric. Sci.,* **6**, 24–38.

Turner, P. D. and Wharton, A. L. (1960) 'Leaf and stem infections of *Theobroma cacao* in West Africa caused by *Phytophthora palmivora*', *Trop. Agric.,* **37**, 321–4.

Turner, P. D. (1960) 'Distribution of strains of *Phytophthora palmivora* from *Theobroma cacao* in West Africa', *Plant Prot. Bull.,* **8**, 53–4.

Turner, P. D. and Asomaning, E. J. A. (1962) 'Root infection of *Theobroma cacao* by *Phytophthora palmivora*', *Trop. Agric.,* **39**, 339–43.

Vernon, A. J. (1966) 'Incidence of Black Pod at Tafo in 1965', *Second FAO Technical Working Party on Cocoa,* Rome 1966. Paper CA/66/28.

Vernon, A. J. (1971) 'Canker: the forgotten disease of cocoa', *Cocoa Growers' Bull.,* **16**, 9–15.

Weststeijn, G. (1968) 'Chemical control of *Phytophthora* pod rot disease', *Ann. Rep. Cocoa Res. Inst. Nigeria 1966–67,* pp. 75–84.

Wharton, A. L. (1962) 'Black pod and minor diseases', in J. B. Wills, ed., *Agriculture and Land Use in Ghana,* Oxford University Press, pp. 333–42.

Wood, G. A. R. (1969) 'Ikiliwindi, Cadbury Brothers Plantation in Cameroon. 2. Black pod disease', *Cocoa Growers' Bull.,* **12**, 9–14.

Wood, G. A. R. (1974) 'Black pod—meteorological factors', in P. H. Gregory, ed., *Phytophthora Disease of Cocoa,* Longman, London 1974, pp. 153–9.

Monilia:

Ampuero, C. E. (1967) '*Monilia* pod-rot of cocoa', *Cocoa Growers' Bull.,* **9**, 15–18.

Desrosiers, R., Van Buchwald, A. and Botanos, C. W. (1955) 'Effect of rainfall on the incidence of *Monilia* pod-rot in Ecuador', *FAO Plant Prot. Bull.,* **3**, 161–4.

Franco, Tamara do (1958) 'Transmission de la Moniliasis del cacao por el *Mecistorhinus tripterus* F.', *Septima Conf. Inter-Amer. de Cacao,* Palmira, Colombia, 1958, pp. 130–6.

Jorgensen, H. (1970) 'Monilia pod-rot of cacao in Ecuador', *Cacao,* **15** (4), 4–13.

Rorer, J. B. (1918) *Enfermadades y plagas del cacao en el Ecuador y metodos modernos, apropiados al cultivo del cacao,* Guayaquil, Ecuador, Asociacion de agricultores.

Witches' Broom:

Holliday, P. (1952) *Witches' Broom Disease of Cacao,* HMSO.

Holliday, P. (1954a) 'Control of witches' broom disease of cacao in Trinidad', *Trop. Agric.,* **31**, 312–17.

Holliday, P. (1954b) 'Spraying against witches' broom disease', *Ann. Rep. Cocoa Res. 1953,* pp. 64–7.

Stahel, G. (1915) '*Marasmius perniciosus* nov. spec.', *Dept. van den Landbouw in Suriname*, Bull. 33, 1915.
Thorold, C. A. (1953b) 'Observations on fungicide control of witches' broom, black-pod and pink disease of *Theobroma cacao*', *Ann. appl. Biol.*, **40** (2), 362–76.

Cushion Gall:
Brunt, A. A. and Wharton, A. L. (1961) 'A gall disease of cocoa (*Theobroma cacao* L.) in Ghana', *Rep. 6th Commonwealth Mycological Conf., Kew, 1960*, pp. 148–56.
Brunt, A. A. and Wharton, A. L. (1962) '*Calonectria rigidiuscula* (Berk. and Br.) Sacc., the cause of a gall disease of cocoa in Ghana', *Nature, Lond.*, **193**, 4818, 903–4.
Hardy, F. (ed.) (1960) *Cacao Manual*, Turrialba, Costa Rica, Inter-Amer. Inst. Agric. Sciences, p. 265.
Hutchins, L. M. (1958) 'Current surveys for cushion gall', *Septima Conf. Interamericana de Cacao*, Palmira, Colombia, 1958, pp. 137–48.
Hutchins, L. M. and Siller, L. R. (1960) 'Cushion gall types in cacao', *Proc. 8th Inter-Amer. Cacao Conf. Trinidad 1960*, pp. 281–9.
Hutchins, L. M., Desrosiers, R. and Martin, E. (1959) *Varietal Susceptibility to Flowery Cushion Gall of Cacao*, Turrialba, Costa Rica, Inter-Amer. Inst. Agric. Sciences, Rep. 33.
Siller, L. R. (1961) 'The relationship between cushion gall and yield', *Cacao*, **6** (3), 6–7.
Snyder, W. C., Thomas, D. L. and Watson, A. G. (In press) 'Fusarium—the unrecognised threat to world cacao production', *Proc. 4th Internat. Cocoa Res. Conf. Trinidad 1972*.
Soria, J. (1960) 'A note on the relationship between flowery cushion gall, self-incompatibility and flower development', *Proc. 8th Inter-Amer. Cacao Conf. Trinidad 1960*, pp. 267–70.

Mealy Pod, Botryodiplodia Pod Rot and Other Pod Diseases:
Dakwa, J. T. (1972) 'Occurrence of mealy pod disease of *Theobroma cacao* in Ghana', *Plant Disease Reporter*, **56** (11), 1011–13.
Legg, J. T. (1970) 'Black pod disease: pod infection', *Ann. Rep. Cocoa Res. Inst. 1968–69*, Tafo, Ghana, pp. 36–7.
Reyes, L. C. de, Reyes, E. H. de and Escobar, F. (In press) 'Etiology of a new pod disease in Venezuela', *Proc. 4th Internat. Cocoa Res. Conf. Trinidad 1972*.
Thrower, L. B. (1960) 'Observations on the diseases of cacao pods in Papua–New Guinea, 1. Fungi associated with mature pods', *Trop. Agric.*, **37**, 111–20.
Turner, P. D. (1968) 'Pod rot of cocoa in Malaya caused by *Phytophthora heveae*', *FAO Plant Prot. Bull.*, **16**, 33.
Waite, B. H. and Salazar, L. G. (1966) 'Pod rot of cacao pods caused by *Fusarium roseum*', *Cacao*, **11** (2), 6–7.
Weststeijn, G. (1966) 'Other pod diseases of cocoa', *Ann. Rep. Cocoa Res. Inst. Nigeria 1964–65*, p. 71.

Ceratostomella Wilt:
Desrosiers, R. (1957) 'Developments in the control of witches' broom Monilia pod rot and Ceratostomella diseases of cacao', *Proc. 6th Conferencia Inter-Amer. du Cacau, Salvador, Bahia, Brazil 1956*, pp. 73–8.
Entwistle, P. F. (1972) *Pests of Cocoa*, Longman, London, p. 631.
Freeman, W. E. (In press) 'Cocoa breeding', *Proc. 4th Internat. Cocoa Res. Conf. Trinidad, 1972*.
Hardy, F., ed. (1960) *Cacao Manual*, Turrialba, Costa Rica, Inter-American Inst. Agric. Sciences, pp. 242–5.
Iton, E. F. (1959) 'Studies on a wilt disease of cacao at River Estate', *Rep. on Cacao Res. 1957–1958*, Trinidad, pp. 55–64.
Iton, E. F. (1961) 'Studies on a wilt disease of cacao at River Estate II. Some aspects of wind transmission', *Rep. on Cacao Res. 1959–1960*, Trinidad, pp. 47–58.

Iton, E. F. (1966) 'Ceratocystis wilt', *Ann. Rep. Cacao Res. 1965*, Regional Res. Centre, Trinidad, pp. 44–50.

Iton, E. F. and Conway, G. R. (1961) 'Studies on a wilt disease of cacao at River Estate, III. Some aspects of the biology and habits of *Xyleborus* spp. and their relation to disease transmission', *Rep. on Cacao Res. 1959–1960*, Trinidad, pp. 59–65.

Saunders, J. L. (1965) 'The *Xyleborus–Ceratocystis* complex of cacao', *Cacao*, **10** (2), 7–13.

Spence, J. A. and Moll, E. R. (1958) 'Preliminary observations on a wilt condition of cocoa', *J. Agric. Soc. Trinidad and Tobago*, **58** (3), 349–59.

Dieback:

Donald, R. G. (1957) 'A capsid control experiment in the Western Region of Nigeria', *Rep. Cocoa Conf., London 1957*, pp. 117–24.

Kay, D. (1961) 'Dieback of cocoa', *Tech. Bull. W. Afr. Cocoa Res. Inst.*, No. 8.

Turner, P. D. (1967) 'Cacao dieback: a review of present knowledge', *FAO Plant Prot. Bull.*, **15**, 81–101.

Vascular Streak Dieback:

Keane, P. J., Flentje, N. J. and Lamb, K. P. (1971) *Vascular Streak Dieback of Cocoa in Papua–New Guinea*, Occasional Papers No. 1, Dept. Biology, University of Papua–New Guinea.

Keane, P. J. and Turner, P. D. (1972) 'Vascular streak dieback of cocoa in western Malaysia', *Cocoa and Coconuts in Malaysia. Proc. Incorp. Soc. Planters Conf., Kuala Lumpur, 1971*, pp. 50–7.

Verticillium Wilt:

Emechebe, A. M. and Leakey, C. L. A. (1968) '*Verticillium dahliae* established as the cause of sudden death of cocoa', *FAO Plant Prot. Bull.*, **16**, 13.

Emechebe, A. M., Leakey, C. L. A. and Banage, W. B. (1971) 'Verticillium wilt of cacao in Uganda: symptoms and establishment of pathogenicity', *Ann. appl. Biol.*, **69**, 223–7.

Leakey, C. L. A. (1965) 'Sudden death disease of cacao in Uganda associated with *Verticillium dahliae* Kleb', *E. Afr. Agric. For. J.*, **31**, 21–4.

Trocme, O. (1972) 'Contribution à l'étude d'une maladie du cacaoyer en Ouganda: le dessèchement eco—fongique des branches', *Café Cacao Thé*, **16**, 219–35.

Pink Disease:

Briton-Jones, H. R. (1934) *The Diseases and Curing of Cacao*. Macmillan.

Edgar, A. T. (1958) *Manual of Rubber Planting*, (Malaya), Kuala Lumpur, Incorp. Soc. Planters.

Thorold, C. A. (1953b) 'Observations on fungicide control of witches' broom, black pod and pink diseases of *Theobroma cacao*', *Ann. appl. Biol.*, **40** (2), 362–76.

Thread Blights:

Leston, D. (1970) 'Incidence of thread blights on cocoa in Ghana', *PANS*, **16** (3), 516–17.

Root Diseases:

Briton-Jones, H. R. (1934) *The Diseases and Curing of Cacao*. Macmillan.

Dade, H. A. (1927) 'Collar crack of cacao (*Armellaria mellea* (Vahl) Fr.)', Dept. Agric. Gold Coast, Bulletin No. 5.

Edgar, A. T. (1958) *Manual of Rubber Planting* (Malaya), Kuala Lumpur, Incorp. Soc. Planters.

John, K. P. (1964) 'Spore dissemination of root disease', *Plrs. Bull. Rubb. Res. Inst., Malaya*, **75**, 233.

162   *Diseases*

Thrower, L. B. (1965) 'Parasitism of cacao by *Fomes noxius* in Papua–New Guinea', *Trop. Agric.*, **42**, 63–7.
Weststeijn, G. (1967) 'Symptomatology and incidence of some root diseases of cocoa in Nigeria', *Nig. Agric. J.*, **4**, 60–3.
Wharton, A. L. (1962) 'Black pod and minor diseases', in J. B. Wills, ed., *Agriculture and Land Use in Ghana*, Oxford University Press, pp. 333–42.

Mistletoes:
Kuijt, J. (1964) 'Critical observations on the parasitism of New World mistletoes', *Can. J. Bot.*, **42**, 1243–78.
Room, P. M. (1971) 'Some physiological aspects of the relationship between cocoa, *Theobroma cacao*, and the mistletoe *Tapinanthus bangwensis* (Engl. and K Krause)', *Ann. Bot.*, **35**, 169–74.
Room, P. M. (1972) 'Mistletoes on West African cocoa', *Cocoa Growers' Bull.*, **18**, 14–18.

Chapter 11

Insects and Cocoa

P. F. Entwistle   *Unit of Invertebrate Virology, Oxford*

Over 1,500 different insects are known to feed on cocoa. Whenever the tree is introduced into a new area it is inevitably attacked by at least some previously unrecorded pests, a process which may continue to some extent even in places where the crop has been long established. Fortunately the effects of the majority of insects is so slight, or epidemics are so ephemeral, that only a handful of species has given cause for long-term concern. The most important of these are the mirids (capsids) which are pests in so many parts of the world, after which leaf-cutting ants in the New World, *Pantorhytes* weevils in Papua–New Guinea and cocoa moth (*Acrocercops cramerella*) in Java and the Philippines must be mentioned.

Some insect species attack only the young tree which having matured is no longer so susceptible. Examples of this which are dealt with below are the cocoa 'bollworm' (*Earias biplaga*), chafer beetles and a shot-hole borer beetle (*Xylosandrus compactus*).

A third category of insects is important mainly as agents of spread of cocoa diseases. The mealybugs which transmit cocoa virus diseases are major examples of this. Others aid disease spread by making cocoa more susceptible, usually because their feeding activities create sites suitable for the entry of pathogenic fungi; for instance shield bugs (*Antiteuchus* species) increase the incidence of *Monilia* pod rot and shot-hole borer beetles (especially *Xyleborus ferrugineus*) of *Ceratostomella* wilt disease.

Chemical control

Although world attention is increasingly being turned to non-chemical methods of pest control, insecticides still remain the mainstay of cocoa protection schemes. Control systems alternative to the purely chemical attempt to lower pest numbers by the combined use of biological control and appropriate cultural methods, with such insecticides as are absolutely necessary being used with minimal prejudice to beneficial animals (insect and otherwise). Such comparatively complex systems, known today as integrated control, will in general be much more difficult to introduce on peasant holdings than on larger commercially run plantations. For this reason the problems of widespread use of insecticides will remain with us for years to come. Accordingly a few general observations are offered on the approach to chemical control of insect pests of cocoa.

In some cocoa-producing countries the use of particular insecticides has official approval signifying that, besides controlling the pest concerned, these compounds are without injurious side effects of an ecological nature and do not taint or cause toxic residues in the final product. But, because this is not always so, particular attention must be drawn to certain hazards which may exist in the use of chemicals for insect control in cocoa. There have been several instances where insects not normally of economic importance have increased to epidemic levels and caused much damage following the use of insecticides either for the control of acknowledged pests or because it seemed a good general prophylactic treatment at the time. For example, the use of dieldrin in Ghana resulted in great increases in numbers of the pod miner (*Marmara*), two wood-boring moths (*Metarbela* and *Eulophonotus*) and a wood-boring beetle (*Tragocephala*). In Sabah a formidable combined spraying programme incorporating endrin, dieldrin, aldrin, BHC and DDT induced increases of the red branch-borer (*Zeuzera*) and of psychid moths (Conway, 1971). Evidence indicates that these contact-acting, largely persistent, compounds had less effect on these pests than on their natural enemies, resulting in a differential kill causing increases in pest numbers. In a proper insecticide evaluation programme this aspect of use will have been considered and the optimal spraying rates and times derived.

A major part of any evaluation programme is to investigate the possibility of causing off-flavours or taint in cocoa beans or of the accumulation there of toxic residues of the insecticide itself or of its breakdown products. Because these are real hazards unapproved insecticides should not be used either on or near bearing cocoa, and instructions for the use of approved compounds should be closely followed, especially with respect to times and rates of application (Gerard, 1969; Hancock, 1968). Certain chemicals cause irreversible inhibition of cholinesterase. Acetylcholine occurs evanescently at nerve endings as a transmitter of impulses. It persists, however, when certain organophosphate insecticides inhibit the enzyme cholinesterase by which it is normally hydrolysed, giving rise to a characteristic gut disorder and bronchial spasm syndrome; death occurs by failure of respiration. Exposure of operators to such chemicals can be fatal, especially if safety precautions are not strictly followed. Clearly such compounds are best completely avoided. Finally certain insecticides, especially organochlorines, decay slowly and can be accumulated in the fat stores of vertebrates. There is little doubt that the wide use of such compounds is a danger to wild life, especially to predatory species at the end of the food chains. But it is less well known that some compounds of relatively low mammalian toxicity can at extremely low dosages kill freshwater fish and crustacea and that, as a consequence, local fisheries can be severely injured. A full list of insecticides is given in Table 11.1 at end of chapter.

Precision in spraying is therefore important. In cocoa with forest tree shade, or with planted shade trees, aerial application of insecticides is difficult for the shade prevents planes flying low enough, much insecticide never reaches the cocoa canopy and there is a large body of spray material free to drift in the higher air speeds pertaining above the cocoa canopy. It cannot be urged too strongly that the application of insecticides be treated responsibly in all ways, having particular regard to the purity of the end product of the crop, the security of the spray operators, the local population and the preservation in its proper ecological balance of the local fauna.

Sap-sucking bugs

Mirids or capsids

By far the most important mirid pests of cocoa are those which occur in West and Central Africa. Here *Sahlbergella singularis* is a pest from Sierra Leone to the Congo Republic while *Distantiella theobroma* is especially active from the Ivory Coast to Western Nigeria. These bugs

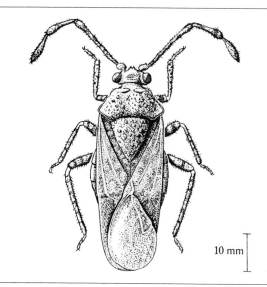

10 mm

Fig. 11.1   *Sahlbergella singularis* Hagl.

have been a major problem in cocoa production for over sixty years. Though not quite so generally important in South and Central America mirids of the genus *Monalonion* have attacked plantation cocoa for over eighty years. The genus *Helopeltis* was first recorded as a pest in Sri Lanka over 100 years ago and has been injurious in Java for many years and also in Malaya. Adaptation of insects to cocoa is a continuing phenomenon and thus, though cocoa has been grown in Papua–New Guinea since before 1905, no mirid attack was seen until the discovery of *Pseudodoniella duni* in 1949; another species, *Pseudodoniella laensis*,

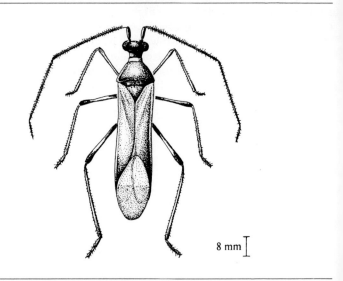

8 mm

Fig. 11.2   *Monalonion annulipes* Sign.

appeared in 1955; the occurrence of *Helopeltis clavifer* on cocoa in Papua–New Guinea dates only from 1954. Similarly cocoa has a long history in Madagascar but mirid attack was unknown until the discovery of *Boxiopsis madagascariensis* on the east coast in 1960. In Sabah, where cocoa cultivation was not seriously undertaken until after 1950, the first mirid, *Platyngomiriodes apiformis*, did not appear until 1963. It seems unlikely we have seen the end of this process of adaptation of unfamiliar mirid species to cocoa.

The life histories of all cocoa mirids are similar. The eggs are buried in the epidermal layer of pods, pod stalks, chupons and fan branches and hatch after ten to seventeen days (possibly less in *Monalonion*). There are then five successive juvenile stages (nymphs) together occupying from eighteen to thirty days, the last of which moults to produce the winged adult insect. The adults are medium-sized (7 to 12 mm long), very slender and with long legs and antennae in *Monalonion* and

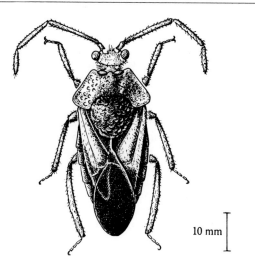

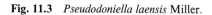

Fig. 11.3 *Pseudodoniella laensis* Miller.

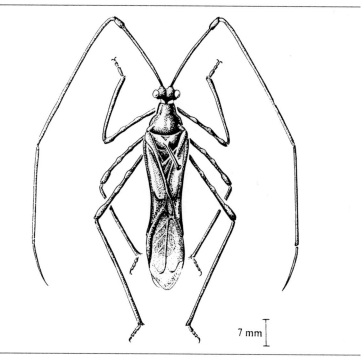

Fig. 11.4 *Helopeltis clavifer* Walk.

especially in *Helopeltis*, but more thick-set in other genera (cf. Figs. 11.1, 11.2, 11.3, and 11.4) (Entwistle and Youdeowei, 1964).

Mirids feed by inserting stylet-like mouthparts into the plant and sucking the juices. The result is small water-soaked areas of tissue, mirid lesions, which rapidly turn black. On pods the lesions are circular but on stems they are usually oval and of somewhat greater size. Both hardened and unhardened stem tissue may be attacked. The direct result of feeding on unhardened stems is wilting and terminal death, but the lesions also serve as the point of entry of injurious fungi. In West Africa this is notably the wound parasite *Calonectria rigidiuscula* which often causes, especially in trees under stress from heavy mirid attack, drought and other adversities, extensive dieback of branches. Elsewhere other organisms may be involved.

The effect of feeding on pods varies regionally. In West Africa, for instance, it is of little importance though cherelles may wilt, and especially heavy feeding may lead to breakdown of the husk and rotting of the bean mass in larger pods. In the Americas and Papua–New Guinea much of the loss caused by mirids is through pod feeding. In general, however, crop loss is the indirect effect of feeding on stem tissue which often results in progressive deterioration of the canopy by dieback of small twigs extending to destruction of branches and, extremely, to death of the whole tree. Naturally this type of damage is very serious for though, as discussed below, trees in an advanced state of attack can be regenerated, some years' crop loss is first experienced.

The pattern of attack is influenced by the cultural conditions under which cocoa is grown. Where there is no overhead shade damage tends to be diffuse and the general scorched appearance which results from many damaged shoots is known as 'capsid blast'. Where there is overhead shade damage tends to be localised below gaps in the overhead shade cover. Here groups of cocoa trees may be extensively damaged while surrounding cocoa may be much less affected, if at all. Such patches of intensive damage are, in West Africa, known as 'capsid pockets'. But, no matter what the status of overhead shade, damage is more frequent where the cocoa canopy itself is incomplete. Degeneration involves the interaction of three main factors: the direct effect of mirid feeding, fungal dieback of stems and competition for water and possibly nutrients from weed growth resulting from increased light intensity at ground level. In parts of West Africa cocoa virus diseases may be an additional degenerative factor. Where overhead shade is absent or inadequate there is the possibility of attack by psyllids and leaf-hoppers contributing to canopy decline.

A knowledge of mirid population cycles is important in timing insecticidal control schemes, for numbers vary greatly during the year. In general throughout the cocoa world, mirid populations decline numerically in periods of low humidity but increase to their highest levels with the abatement of the main rains; very heavy rain itself appears to depress their numbers, though not as greatly as does low humidity (Gibbs *et al.*, 1968; Lavabre *et al.*, 1963).

Mirid control In West Africa extensive and satisfactory control of mirids has for years been achieved with gamma-BHC to which the cocoa industry has indeed owed much of its stability. It is normally applied at the rate of 140 to 280 g of active ingredient per hectare in 22·5 to 45 litres of water (2 to 4 oz per acre in 5 to 10 gallons), depending on the mode of application.

However, in 1961 resistance to BHC was found in *Distantiella theobroma* in parts of Ghana and in 1962 in *Sahlbergella singularis* in Western Nigeria. Since then the areas occupied by resistant mirids in these two countries have increased and within them adequate control with BHC cannot be expected. Resistance to BHC also confers on the mirid resistance to other cyclodiene organochlorine insecticides such as aldrin, dieldrin, endrin and heptachlor. From the search for suitable alternative, but chemically unrelated, insecticides the carbamates propoxur (arprocarb) and Orthobux have emerged as effective miricides. At 140 g and 280 g, respectively, per hectare (2 oz and 4 oz per acre), mistblower-dispersed, they seem satisfactory both on grounds of efficiency and general safety. It must be emphasised, however, that at the time of writing these results are provisional in that they have not yet been officially adopted (Collingwood, 1971).

In Ghana the recommended spraying practice has been for two double applications of BHC each year in each of which the two sprays are at twenty-eight day intervals, in June–July and November–December. The theory behind this system is that the first spray of each pair destroys nymphs and adults, and the second kills any mirids which have emerged from eggs present at the first spraying, before they can mature reproductively. The June–July applications are timed to coincide with the beginnings of population increase; those of November–December cover the period of maximum population development. However, recent work on the timing of spray applications in Ghana has indicated that this does not give as good control as does spraying in August, September, October and December. This is very close to the system which has for some years had official backing in Western Nigeria which is for three applications of BHC at intervals of twenty-eight days beginning in August and for spot treatment to deal with residual foci of infestation.

The advent of comparatively low-volume spraying techniques and the development of easily portable spraying machines has allowed the wide-scale treatment of cocoa with insecticides where previously shortage of water and roughness or remoteness of terrain was limiting. Knapsack mistblowers powered by two- or four-stroke motors allow fairly rapid application of insecticides at volumes as low as 56 litres per hectare (5 gallons per acre) and have a distance of throw adequate to place spray droplets in or near the tops of mature cocoa trees. Pneumatic knapsack machines have a much less powerful throw but when fitted with a long spray lance can give good results (Higgins, 1964, 1965). In Nigeria they fulfil the dual purpose of mirid and black pod disease control. It should be emphasised that with volatile insecticides like BHC and propoxur the fumigant action of the insecticide tends to compensate for inequalities in spray application. When less volatile insecticides are used greater attention has to be paid to obtaining good coverage of trees with the spray. Thus adequate mirid control was achieved in Nigeria by applying BHC in only 11 litres (2·5 gallons) per 100 trees, while non-volatile (but nevertheless highly active) chemicals had to be applied in 22·5 litres (5 gallons) to achieve the same level of control.

Much practical control work has also been done in West Africa, especially in French-speaking areas, with portable fogging machines in which the toxicant is disseminated as a cloud in diesel oil. Its success depends very much on employment of a volatile insecticide and on treating either early or late in the day when convection currents are slight and the fog lingers longest in the cocoa canopy (Mire, 1965).

In the Lukolela plantations of the Congo Republic, where it was the practice to keep the height of cocoa low by arresting growth at the first jorquette, mirid control was practised as part of a routine monthly inspection cycle in which any mirid-damaged shoots were pruned off. The affected trees were then treated with a 2·2 per cent BHC dust which, because of the low stature of the trees, could be efficiently administered with an inexpensive hand-bellows duster (Nicol and Taylor, 1954).

Much has been written about the use of ants to control mirids on cocoa. Though the Black Ant, *Dolichoderus bituberculatus*, was once used to deter *Helopeltis* in Java it is doubtful if this practice now continues (Meer Mohr, 1927). However, it is interesting to note that to drive out mirids farmers in parts of Cameroon artificially infest trees with the ant *Wasmannia auropunctata* (de Mire, 1969). In Ghana and Nigeria a lot of attention has been paid to the relationships of the dominant ants and the two most important mirid species. The numbers of *Sahlbergella singularis* are not depressed by ants and may even be slightly increased, perhaps because ants are inimical to some enemies of mirids. On the other hand, numbers of *Distantiella theobroma* are depressed by the aggressive ant *Oecophylla longinoda*. The extent of protection which it affords is determined by its prevalence; it seldom infests more than 20 per cent of trees and usually less. *Oecophylla* is not likely to become a primary biological control factor but it should certainly be considered as an important component in integrated control of mirids. In Papua–New Guinea where at present the ant *Anoplolepis longipes* protects cocoa from mirid attack, not by predation but by disturbance as does *Dolichoderus* in Java.

It is possible to encourage even quite seriously damaged cocoa to regenerate and to resume a satisfactory level of crop production. This is brought about primarily by regular and efficient chemical control of mirids and other pests. But in addition certain cultural operations must be performed. Seedlings of a vigorous variety should be planted in vacant stands, but the gaps in the canopy of the cocoa must be temporarily made good with banana or plantain which should be removed as the canopy closes. Where there is overhead shade the canopy of which has been damaged, this must be made good. Badly affected areas should be weeded regularly, but as the canopy regenerates this will gradually become unnecessary. It has been found in Nigeria that even cocoa affected by local virus diseases, in addition to mirids and *Calonectria*, may be regenerated to give economically acceptable yields. In Ghana, however, where some of the most virulent viruses occur, this would often be impossible.

Shield bugs

Antiteuchus (= Mecistorhinus). Attack occurs over a wide area of South and Central America and Trinidad. There is no precise information on the extent of the directly adverse effects of feeding activities, though in Trinidad both *Antiteuchus picea* and *A. tripterus* may be important as minor pests in nurseries and propagators (Callan, 1944). This aspect has been overshadowed by their implication in increasing the incidence of watery pod rot (caused by the fungus *Monilia roreri*) which is said to gain entrance through feeding scars. In three separate experiments comparing the incidence of pod rot in pods inoculated with *Monilia* spores in the presence and absence of *A. tripterus*, pod disease increased from 35 to 76, 3–15 to 77 and 28 to 62 per cent where the insect was present. But definitive experiments on the insecticidal control of this bug during periods of greatest pod 'infection risk' have not been reported. *A. tripterus* eggs are deposited on all parts of the tree and even on fallen branches and leaves; it is a shade loving insect especially remarkable because it tends its eggs and also its larvae whilst these are very young so, outside the bees, wasps and ants, providing one of the rare instances of maternal care in insects (Sepulveda, 1955; Franco, 1958).

Bathycoelia This large green shield bug (adult 20 mm long; Fig. 11.5) attacks cocoa from the Ivory Coast to the Congo Republic. The eggs of *Bathycoelia thalassina*, the most important species in Ghana and Nigeria, are mainly laid on leaves but also on trunks and branches. Feeding, however, is primarily on pods and crop loss is by inhibition of bean development and pod abortion. The long feeding stylets penetrate the pod husk and suck out the contents of the beans, so that some become brown whilst others never mature, leaving only a microscopic lesion on the pod surface. Young pods generally turn yellow, then

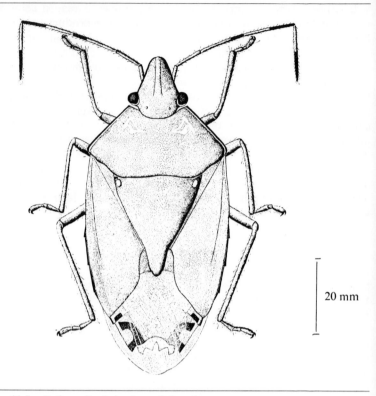

20 mm

Fig. 11.5   *Bathycoelia thalassina*, adult male.

black, but larger pods stop growing and become yellow, a condition
called 'premature ripening'. They may turn yellow basally or distally
depending on where the feeding punctures are made. Their colour
later becomes brown and then black (Lodos, 1967a).

Feeding being so much restricted to pods the development of large
populations is possible only on cocoa in which fruit are produced
throughout the year. This largely excepts Amelonado and indeed it
was not until hybrid and Amazon plantings became common that
Bathycoelia emerged as a problem.

Evaluation of insecticides in Ghana has, due to the resistance of
cocoa mirids to cyclodiene compounds, excluded organochlorines,
except DDT, as possible control agents. The most promising compounds for mirid control are insufficiently toxic to *Bathycoelia*. The
best field results have been with compounds which unfortunately cause
high and irreversible levels of cholinesterase inhibition in man and
consequently satisfactory control recommendations are not available
at present (Marchart and Lodos, 1969). However trials in Cameroon

showed that endosulfan (not tested in Ghana) gave good control and that fenitrothion was also effective, though it did not show up too well in the Ghana trials.

Coreid bugs In Africa members of two closely allied genera, *Theraptus* and *Pseudotheraptus*, are especially associated with pods. *Pseudotheraptus devastans* feeding inhibits development of cherelles and causes distortion of older pods. It is especially prevalent on hybrid and Amazon cocoa probably because pods are more continuously present throughout the year than on Amelonado cocoa (Lodos, 1967b). Generally similar to these bugs is *Amblypelta*. In the Solomon Islands *Amblypelta coccophaga* attacks only stems but in Papua–New Guinea *A. theobromae* (Fig. 11.6) feeds mainly on cherelles and young pods causing distortion and necrosis (Brown, 1958).

Leaf hoppers, psyllids and aphids

Many species of these sap-sucking Homoptera occur on cocoa, but

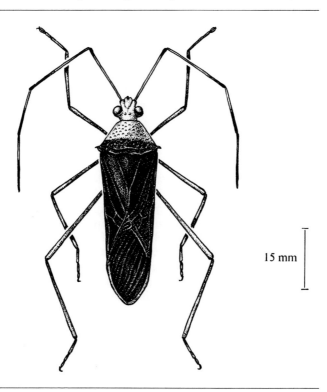

15 mm

Fig. 11.6 *Amblypelta theobromae* Brown.

most are of minor importance. Among the leaf hoppers *Empoasca devastans* (Fig. 11.7) in Sri Lanka (Fernando, 1959), *Affroccidens* species (Typhlocybidae) in Ghana (Lodos, 1969) and *Chinaia rubescens* (Coelidiidae) in Costa Rica cause distortion and premature fall of leaves (Salas and Hansen, 1963). Such damage has been referred to as 'leaf hopper burn' and may at times be an important factor in canopy degradation. In Brazil, Guyana, Colombia, Costa Rica and Trinidad *Horiola picta* (Membracidae), a species much attended by ants, feeds on flower cushions, pods and stems and may cause pod wilt.

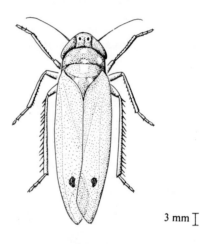

3 mm

Fig. 11.7   *Empoasca devastans* Dist.

Cocoa psyllids are restricted to Africa where *Tyora (Mesohomatoma) tessmanni* predominates from Sierra Leone to the Congo Republic. Adults (only 4 mm long; Fig. 11.8) lay eggs in vegetative buds, developing leaves, petiolar swellings of leaves and also flowers and very young fruit (Cotterell, 1943). Flowers and pods may atrophy following attack, while under conditions of drought and high insolation deposition of large numbers of eggs in terminal buds results in desiccation and death of the bud with growth retardation of the shoot. Thus psyllids may be regionally potent factors in canopy degradation. Control methods have not yet been developed.

Aphids (especially *Toxoptera aurantii*), are seldom serious pests but leaf crinkling, premature leaf fall and flower wilt occur, while young stems may wither and affected plants etiolate (Kirkpatrick, 1955). Exceptional attack occurred on potted seedlings in Uganda following DDT spraying against leaf-eating caterpillars. The seedlings were severely distorted but the infestation was controlled with 0·25 per cent menazon.

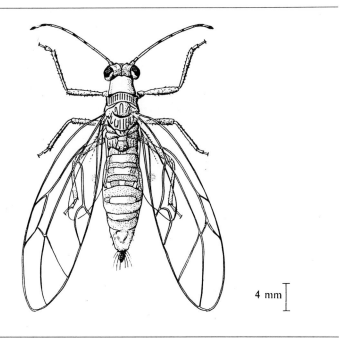

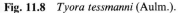

4 mm

Fig. 11.8 *Tyora tessmanni* (Aulm.).

Mealybugs and scale insects

These are small usually flattened, scale-like insects with long stylet-like mouthparts with which they suck sap. The dorsal surface is generally covered with a hard scale or, in the case of mealybugs and some others, with white mealy secretion. Biologically the group is particularly unusual because the adult female is wingless and has a juvenile structure. Hence distribution takes place by the young nymphs being carried by air currents. They excrete honeydew for which they are often closely attended by ants, some species of which construct protective tents of soil and vegetable matter over groups or colonies of bugs (Entwistle, 1972, 1973).

In any one area there is generally a complex of species on cocoa and no cocoa-growing region is free from them. They are seldom frequent enough to cause serious damage but sporadic examples of severe attack are not uncommon. Mealybugs, however, are the only known carriers of cocoa virus diseases and so are of especial importance in West Africa where the ravages of swollen shoot disease and of allied viruses have cost many millions of pounds in crop lost and in control measures. Their suppression, in so far as this is practicable, is discussed under the heading of 'Virus diseases' in chapter 10.

Cocoa thrips

A number of different thrips attack cocoa but by far the most frequent to do so is *Selenothrips rubrocinctus*, the cocoa or red-banded thrips. This is a very small (adult 1·5 mm long) elongate insect, black as an adult but in the juvenile stages pale yellow with a red basal abdominal band (Fig. 11.9). It occurs on cocoa and many other plants throughout the tropics, but has particularly attracted attention in the West Indies, Surinam (Reyne, 1921) and São Tomé (Cotterell, 1930).

Eggs are inserted beneath the lower epidermis of the leaf. Sap-sucking, which causes damage to the leaf cells, results in the leaf becoming silvered.

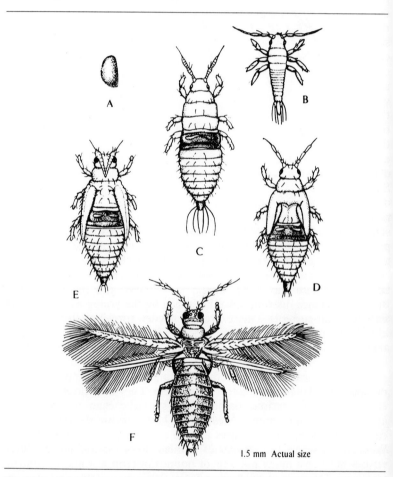

1.5 mm Actual size

Fig. 11.9 *Selenothrips rubrocinctus* (Giard). (A) egg; (B) and (C) first and second instar nymphs; (D) prepupa; (E) pupa; (F) adult female.

The nymphs carry the abdomen curved upwards, with a drop of clear fluid poised on the hairs at the apex. This is periodically released and drops on the leaf surface where it dries to form a brown spot. The speckling caused by the presence of many such dots on the partially dried or silvered tissue is characteristic of thrips injury (Fennah, 1947).

For a long time thrips were considered to be a pest of primary importance. However, the work of Fennah in Trinidad has shown that the establishment of thrips on cocoa leaves is only made possible by the indirect effect of adverse factors in the physical and nutritional environment deranging normal metabolism of the leaf. Thrips are presumed able to 'recognise' the existence of such a state which is conducive to their multiplication. It is typical of cocoa thrips attack that some areas are persistently more attacked than others, often outside the main season of thrips abundance, and it has been shown certain conditions of the soil, e.g. persistent waterlogging, may give rise to this circumstance (Cotterell, 1928). Conditions suitable to attack may also be created. This happened in the West African island of São Tomé where the great fall in production in the 1920s was attributed to the rise of thrips as a pest; the latter was coincident with the deliberate destruction of overhead shade and of protective windbreaks and erosion on the many steep slopes on which cocoa was grown. Further supporting evidence that cocoa thrips cannot be a primary pest lies in the fact that chemical control has generally failed to induce the desired response. Such beneficial responses as have occurred are associated with the use of the fungicidal Bordeaux mixture and can be readily interpreted as resulting from the incidental supply of the need for small quantities of some trace metals that were the cause, or contributory to it, of the protein-synthesis derangement which rendered trees susceptible to thrips in the first place (Fennah, 1955, 1963, 1965).

However, the bronzing of the epidermis of pods makes the distinction between ripe and unripe pods difficult with the consequent premature harvesting of some beans and this may cause a decline in the standard of the final product. As a temporary expedient in such situations thrips may be controlled by spraying BHC, but the long term answer lies in correcting adverse cultural conditions.

Caterpillars

Ring bark borers

Endoclyta (= Phassus) hosei and *Phassus sericeus (= damor)* are ring bark borers in Sabah (Conway, 1971) and Java (Kalshoven, 1922), respectively. The vernacular name arises from the habit of the larva of devouring the bark around the stem and this is done beneath the cover of a web of silk and bark particles; also a deeper escape tunnel is excavated into the wood. *Endoclyta* usually attacks trees six months to two and a half years old at the collar and older trees at the jorquettes and upper branch unions. Control can be achieved by squirting 1 per cent dieldrin emulsion into holes and sealing them with wet earth. This is done on monthly inspection cycles, the regularity of which during the first three years after planting out is important.

Trema cannabina v. *glabrescens* appears to be the primary host plant and grows in profusion following bush clearance. Its eradication contributes considerably to control.

Larvae of *Phassus* first enter dead and more or less rotted twigs on the ground but soon move to nearby living plants, a preference being shown for slender stems 1 to 6 cm in diameter. Attack is mainly at the collar region and may result in a ridge of wound tissue on the upper side of the gallery, but this is often concealed by debris on the surface of the soil.

Red branch borer

Zeuzera coffeae (Fig. 11.10) has been recorded as a pest of cocoa from

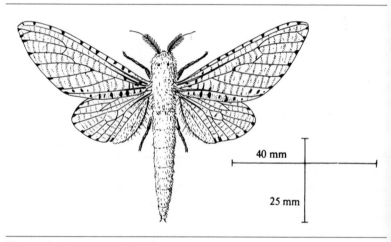

40 mm

25 mm

Fig. 11.10   *Zeuzera coffeae* Nietn.

Sri Lanka, Malaya, Java (Kalshoven, 1919) and Papua–New Guinea, and it is suspected that attack in Sabah is by both this species and *Z. roricyanea* (Conway, 1971). The adult is a leopard moth with a wing span of about 4 cm. Young larvae disperse on silk thread 'parachutes' and then burrow into the phloem and cambium of suitable woody plants. Damage is especially serious in young cocoa, seedlings of which may be destroyed and slender branches may dry out and often snap off. The larva tunnels for some distance (9 up to 30 cm) along the centre of slender stems and finally makes a transverse tunnel before pupation. The total life cycle is probably four to five months (Kalshoven, 1940).

Control seems best achieved, and maintained, by pruning off and destroying attacked branches possibly augmented, on trunks and larger branches which it is inadvisable to remove, by application of DDT or dieldrin to the immediate vicinity of galleries. General application of persistent insecticides can, however, lead to increases in *Zeuzera* populations.

Cocoa moth

Larvae of this small moth, *Acrocercops cramerella* (Fig. 11.11), are

12·5 mm
5 mm

Fig. 11.11 *Acrocercops cramerella* (Snell.).

extremely serious pests of pods in Java and the Philippines. It is also known from the Celebes and Papua–New Guinea. The eggs are laid on the epidermis of pods, preferably in the furrows, and hatch in six to nine days. The tiny caterpillars bore straight through the husk and make long galleries filled with excrement between the beans. When fully grown after fifteen to eighteen days (10 to 12 mm long), they leave the galleries to pupate in a flat membranous cocoon on the pod surface, fallen leaves or weeds. The pupal period is five to eight days. Damage is scarcely detectable from the outside of the pod but the beans though not actually attacked are rendered worthless.

Rampassen is the only successful control method yet developed. It consists of removal of all pods from a plantation at the end of each main crop period in order to break the sequence of generations of the pest. It is essential that all pods removed be destroyed and that alternative host plants, especially *Nephelium* (rambutan), *Cola* and *Cynometra cauliflora*, be also removed (Roepke, 1912). Cocoa of Criollo type seems more susceptible than Forastero and this may be due to either one or both of two factors: first, the woody sclerotic layer of the pod is poorly developed in Criollo and so larvae penetrate easily to the beans, and secondly the deeper ridging of the Criollo pod may favour oviposition by cocoa moth. Control with endrin or DDT is possible, but at the cost of repeated applications (Laoh, 1954).

Cocoa 'bollworm'

This smallish green moth, *Earias biplaga*, is found throughout Africa south of the Sahara and is best known, from the form of the caterpillar, as a spiny bollworm of cotton. Cocoa is attacked from the Ivory Coast to the Congo Republic, at least, and in São Tomé. Attack is mainly confined to plants up to three years old and is most intense when there is no overhead shade. Because of this it has become particularly noticeable following planting on clear-felled land, especially when there is inadequate provision of nurse shade. Young larvae destroy apical buds and older ones mainly feed on flush leaves. The moth is present throughout the year so that continuous disbudding is possible, leading to extremely malformed plants in which canopy formation is delayed and may be prevented. In Ghana removal of nurse shade two years after planting resulted in heavy attack by *Earias* and in an associated, undiagnosed, stem dieback.

No satisfactory chemical control regime has been developed (Nguyen-Ban, 1971), and there is no doubt that avoidance of *Earias* attack depends on the provision of adequate shade for plants up to at least three years old in the field. Heavy natural parasitism has not prevented a serious level of damage (Entwistle, 1969, 1972).

Cocoa armyworm

A wide ranging species (*Tiracola plagiata*), India to Australia, which has only been found attacking cocoa in New Guinea. Only occasional local infestations were known until epidemic populations developed, particularly in the Popondetta area. Extensive tracts of virgin forest had been clear-felled and burnt over and many weed species swiftly became established, some of which became heavily infested (these included *Erechthitis hieraciifolia* and *Euphorbia cyathophora*). The *Leucaena leucocephala* and *Crotalaria anagyroides* used as shade were unfortunately also acceptable hosts on which populations built up to enormous proportions and when cocoa was planted it was very soon infested.

57. (*left*) Pods attacked by *Helopeltis*, Nigeria.

58. Cocoa farm devastated by capsids. A tree on the right shows stag-headed appearance.

59. Cocoa farm in Ghana suffering from capsid blast.

60. (*right*) Harvesting on a plantation in Came-

61. Opening pods, Ghana. A cutlass is used to split the pods and scoop out the beans.

62. (*right*) Opening pods, Cameroon. The pods are broken on a concrete block; the empty husks are thrown into hoppers on the left.

64. A heap uncovered. In this case the placentae have not been removed—this is not the usual practice.

63. A fermenting heap, Nigeria.

Selective feeding on the flush and growing points of cocoa leads to large scale destruction of apical dominance. Few trees die but canopy development is severely prejudiced, the age of bearing is delayed and the trees become misshapen with much upward spindly growth (Catley, 1963a; Dun, 1967).

Adequate control has been reported by spraying at high volume (80 to 120 litres per ha) with 0·25 per cent DDT, 0·2 per cent carbaryl or 0·05 per cent endrin (Catley, 1963b). As this needs to be repeated frequently its economics must be questioned and a compromise between less spraying and some reduction in *Leucaena* shade should probably be considered.

Ants

Leaf-cutting ants

These are restricted to the New World where the principal species, *Atta cephalotes*, is generally distributed in South and Central America

Fig. 11.12   Leaf damage by workers of *Atta cephalotes* (L.), Brazil.

and in Trinidad. In the latter island, Brazil and Costa Rica species of *Acromyrmex* may also be important. *Atta* is indisputably the most important genus in South America but in Trinidad where land is cleared of forest and planted to citrus or used for general agricultural purposes, nests of *Acromyrmex octospinosus* are very much more common. (In Brazil *A. cephalotes* is known as 'suava de mata', in Costa Rica as 'zampopas' and in Trinidad, with *A. octospinosus*, as 'bachac ants'.)

Workers of leaf-cutting ants bite oval pieces from the leaves of many trees, including cocoa (Fig. 11.12), on which in 'gardens' in their sub-terranean nests they cultivate the fungus which is their principal food. Flowers, cherelles and the surface of pods may also be used. To ensure the future of each new colony the virgin queen before her mating flight stores a small portion of the fungal material in a pouch inside her mouth. The nests (Fig. 11.13) may become very large, exceptionally covering a quarter of a hectare (0·6 acre). Those of *Acromyrmex* are much smaller, a metre or less across.

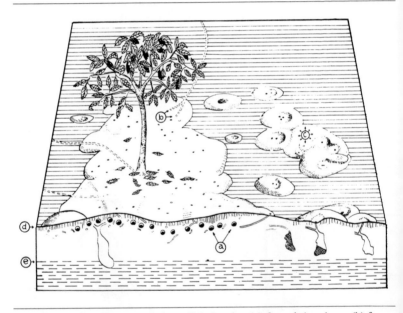

Fig. 11.13   Nest of *Atta cephalotes* (L.) showing (a) fungal chambers; (b) forag-ing tracks; (c) nest entrances; (d) ground level and (e) water table.

Foraging workers of *A. cephalotes* may range up to 100 to 200 metres in search of suitable plant material and foraging up to 0·8 km has been recorded (Cherrett, 1968).

The total losses (all crops together) from leaf-cutting ants has been estimated at £400 million, which as pests puts them on a plane with locusts in the Old World. Losses from 'bachacs' in Trinidad have been calculated at 3·4 per cent of the cocoa crop annually or a total of £55,000, but this is thought to be an underestimate as it is based on the cost of control and does not take crop losses into account. But if it can be assumed this is a minimal loss throughout American cocoa then, on the basis of production for 1966–69, an annual loss of 10,000 long tons may be suggested (Cherrett and Sims, 1968).

Nests may be eliminated by introducing insecticides as dusts into their entrances. The dosage rate is calculated on the surface area of the nest; 22 g per m^2 of a 5 per cent heptachlor or dieldrin dust gives acceptable control. Fumigation with methyl bromide has also been widely used in Brazil. In Costa Rica physical destruction of the nests of *Atta cephalotes* and *A. colombica*, followed by application of chlordane, is compulsory under national law.

Eradication of ants' nests in forest, scrub or other rough land adjacent to plantations may be very difficult. In situations like this baits can be effective and J. M. Cherrett has investigated the control of 'bachacs' in Trinidad using as a bait citrus meal impregnated with mirex, a slow acting stomach poison. Silicone waterproofing increases the water repellency of citrus meal bait so that it retains its attraction longer. Because of the low dosage rate (provisional data) and because it seems, though the point requires verification, that there will be few ecologically adverse side effects, this formulation appears ideally suited to aerial distribution. Provisionally a dosage rate of 10 g mirex in 2·3 kg of bait per ha (0·14 oz in 2 lb per acre) has been suggested (Cherrett, 1969; Cherrett and Merrett, 1969; Cherrett and Sims, 1969).

Enxerto, Cacarema and Balata ants

Azteca is another New World group of ants which, unlike leaf-cutting ants, has the habit of tending plant sap-sucking bugs (aphids, whitefly, scale insects, plant hoppers, etc.) for the sake of their honeydew excretions. Association with cocoa is by no means obligatory and the ants are to be found on many woody plants, often on the shade trees in cocoa plantations. Some build carton nests on aerial parts of trees, whilst others form their nests in hollow places; *Azteca velox* in Trinidad, for instance, makes nests in the old tunnels of cocoa beetle (*Steirastoma breve*).

A. paraensis v. *bondari*, the 'Enxerto Ant' of Brazil, builds a spherical nest which in the course of time comes to be held together by the roots of epiphytic plants—orchids and others. Its food is mainly honeydew, but it is directly injurious to the tree because it bites young terminal shoots to obtain mucilage for nest building. This leads to terminal defoliation, loss of apical dominance and the growth of many less vigorous shoots giving a broomlike appearance. Below the broom the leaves arise in a cluster resulting in a very characteristic picture enabling detection of attack from a distance (Silva, 1957).

A. chartifex in Trinidad and its variety *spiriti* in Brazil, where it is known as the 'Cacarema ant', make pendulous nests on sloping trunks and branches. Trees infested with *Azteca* ants in Trinidad, where they are commonly known as 'Balata ants', are said never to do so well and to have pods which are dwarfed and generally disfigured by scars and brown patches.

Control is one of the major entomological problems of the Brazilian cocoa industry; there may be 150 colonies per ha. In the mid-1950s it was estimated that cocoa plantations in Brazil, including their shade trees, harboured 20 to 25 million nests which annually caused losses of £5 m.

The pest can be controlled quite efficiently and inexpensively by injection of 1 per cent BHC dust into the nest. This is best done by a small conventional hand pump to which is attached a tube perforated with a series of holes towards the end and having a sharp tip. The perforated region is introduced into the nest and dust injected by the plunger. To give good distribution the probe is inserted at four separate points and a total of about 15 g of dust administered (Silva and Bastos, 1965).

Beetle pests

Rose beetle and other chafers

The larvae of chafer beetles (Scarabaeidae) feed on roots and decaying organic matter, especially in the soil. Adults eat soft leaves and flowers and it is mainly in this capacity that they are pests of cocoa. Adults generally feed at night. A complex of species damages cocoa in Malaya (*Apogonia*, *Anomala* and *Chaetadoretus* species) and probably elsewhere in the Far East (Lever, 1953). In Java, and especially in Fiji and Samoa, *Adoretus versutus*, Rose Beetle (Fig. 11.14), is a serious defoliator. Young plants are particularly susceptible to attack but may be given physical protection by individual split bamboo or palm leaf fences or, as used more recently in Fiji, by cylinders of plastic gauze. The fences should be a little taller than the cocoa. As adults often hide in soil and ground litter by day the application of chlorinated hydrocarbon insecticides persistent in soil, e.g. BHC, has been recommended. The use of dieldrin in this way is inadvisable unless on a very small scale. Treatment of foliage with lead arsenate or trichlorphon, both stomach poisons, should give control of feeding adults.

In a different category as a pest is *Camenta obesa* which is said to have reduced the area under cocoa in the West African island of Fernando Po (Cotterell, 1930). Both larvae and adults feed on the principal roots of cocoa trees. One or two larvae can kill a two-year-old

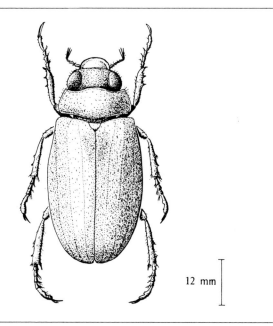

12 mm

Fig. 11.14 *Adoretus versutus* Har.

tree, but mature trees are much more tolerant. Pouring 'coal tar derivatives' or BHC (1·5 per cent) into holes in the soil inclined towards the tap root is said to give good control (Nosti, 1953).

Cocoa beetle

This longhorn beetle, *Steirastoma breve*, Cerambycidae (Fig. 11.15), is known only from the New World where it occurs from the Argentine to Florida. Within this wide range it is a pest of cocoa only in the Guianas, Venezuela, Colombia, Ecuador, Trinidad, Grenada, Martinique and Guadeloupe, and apparently not in Brazil or Central America. The length of the adult varies from 12 to 30 mm and the colour from dark grey to blackish. The adult female bites holes in the bark in each of which she then lays an egg, afterwards closing the holes with her mandibles. The eggs hatch after four or five days and the yellowish larvae bore in the bark where they feed on the cambial layer and the bark itself. At first a rounded chamber is made about the point of oviposition and this is later increased in size and elongated until it forms a tunnel. Externally the larval tunnels are betrayed by a gummy, gelatinous exudation which escapes by small holes made by the larvae through the bark. The larva generally tunnels in a spiral so that young branches and stems often become entirely ringed and die. The mature

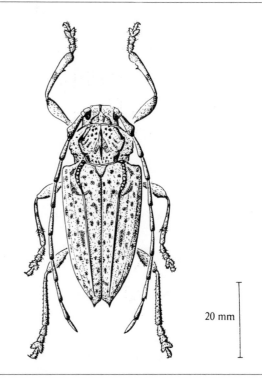

20 mm

Fig. 11.15   *Steirastoma breve* (Sulzer). Adult male.

larva measures 35 to 40 mm long and the larval period lasts two or more months. A pupal chamber opening to an oval exit hole is then excavated in the heart wood and this may considerably weaken branches and young stems. The pupal period is eight to twelve days and the adults rest a few days in the wood before emerging. Hence the life cycle may be as little as three months but can be extended considerably beyond this. There may be three to four generations a year but most egg-laying occurs in the dry season so that there are two main broods annually.

Attack is mainly on trees six months to five years old. Most eggs are laid at the collar, branch forks and wounds of various sorts, and wounds resulting from pruning or bud grafting are very susceptible. Creviced bark, whether in the region of cut branches or not, is favoured. Trees subject to greater light intensity have been found to have higher larval numbers and this has been attributed to the fact that, due to greater photosynthetic activity, the bark of the young stems on which adults feed is more nutritious (Guppy, 1911; Urich, 1925).

In the past control depended on two non-chemical methods: collection of adults and larvae and trapping. The collection of larvae has been shown to be ineffective if not positively injurious because of the physical damage caused in cutting them out. Trapping depends on the especially attractive qualities of *Pachira insignis* (called 'chataigne maron' in Trinidad), one of several wild host plants, in which adult beetles lay eggs very freely. Bundles or piles of *Pachira* wood are placed near to susceptible cocoa and after three weeks (wet season) and two weeks (dry season) the old trap wood is destroyed and replaced with fresh. This method has probably fallen into disuse.

If applied with a sticker some organo-chlorine insecticides may give better control than lead arsenate. However, lead arsenate acts as a stomach poison and so does not affect the natural enemies of cocoa beetle or of other insects attacking cocoa. The widespread use of persistent organo-chlorine insecticides can cause increases in the numbers of some cocoa pest species, but this is unlikely to be brought about by lead arsenate. One available recommendation is to apply lead arsenate at 2 kg emulsified in 200 litres of water by 700 ml of linseed oil emulsion (made by adding 'Teepol', Triton B 1956, or some similar wetting agent, to raw linseed oil) per hectare (1·8 lb in 18 gallons of water by 10 fl oz of linseed oil per acre). Trees should be sprayed where adult beetles feed, that is on the younger, smooth-barked, branches and shoots (Fennah, 1948, 1954).

Glenea longhorn borers

The longhorn beetle genus *Glenea* (Cerambycidae) is widespread in the Old World tropics but pest species are known only in Java, Papua–New Guinea and New Britain, though slight attack has been recorded from Malaya. Elsewhere several species favour dead and perhaps dying trees.

Attack seems specially common in neglected, overgrown plantations. For instance, *Glenea aluensis*, which thrives best in heavily shaded situations, was able to gain a strong hold in the Gazelle Peninsula of New Britain when plantations became overgrown during the war. Similarly *G. lefebueri* in New Guinea is not normally a problem in well-maintained plantations. In Java *G. novemguttata* seems to occur especially in plantations bordering on the original jungle.

Eggs are deposited deep in the bark and larvae burrow below the bark making at intervals small holes through which frass and mucilage seep. When full grown, in about two to three months, each excavates an oblong pupal cavity in the wood. The total length of the larval gallery is 10 to 20 cm and it may nearly girdle stems of modest diameter. Multiple infestations may result in girdling of stems and branches of mature trees, causing death. Trees become susceptible to *G. lefebueri* in the third year and attack is concentrated on the lower trunk (Dun, 1951; Schreurs, 1965).

Control is reported to be possible by excising larvae on a three to four week inspection cycle; at longer intervals difficulty arises in extracting more deeply embedded, older larvae. One per cent dieldrin has given 90 per cent control, as measured at one week after application, but until further tests have been carried out its use cannot be recommended unless on a small scale (Schreurs, 1965). Use may perhaps be made of lead arsenate, for the feeding habits of the adult are similar to those of 'cocoa beetle' in the New World.

Weevil borers

The economically very important genus *Pantorhytes* is restricted to Papua–New Guinea, the Bismarck and Solomon islands, with one species extending into the Cape York Peninsula of Australia. At least six species attack cocoa (Gressitt, 1966).

Since the previous edition of this book attack by *Pantorhytes* weevils (Curculionidae) have loomed large in the extensive post-1960 plantings in the Popondetta area of Northern District, Papua–New Guinea. Here *Pantorhytes szentivanyi* (Fig. 11.16) has become the most significant of the long-term pests. But increasing importance has been noted in the Morobe and Milne Bay Districts and in New Britain. *P. biplagiatus* is a pest in the Solomon Islands.

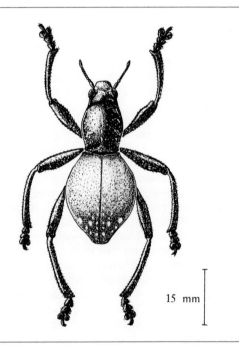

15 mm

Fig. 11.16　*Panytorhytes szentivanyi* Mshl.

Eggs are laid in crevices in the bark, especially at the jorquette and branch unions, but later on the trunk and branches themselves. Larvae (Fig. 11.17) burrow in the stem or branches to a depth of about 1·0 to 1·5 cm and feed in tunnels more or less parallel to the surface. The effect of many larvae feeding round the jorquette is to cause cracking of the stem leading to death of the tree. Mechanical damage following wind is a feature of infested plantings and affected trees may split at the jorquette. Ring-barking may also result. When populations are very high larvae may occur in pods.

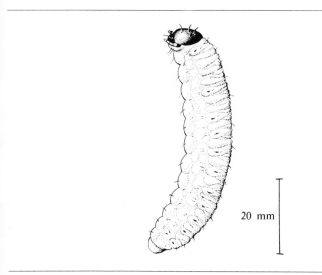

20 mm

Fig. 11.17   Larva of *Pantorhytes proximus* Fst.

Adults are large (15 mm long) (Fig. 11.16), flightless, and tremendously long lived, up to two years for *P. szentivanyi*. Adults feed on young leaves, the veins of old leaves, on the bark of shoots up to six months old and the husk of pods where they leave oval scars about 1·5 cm in area. They also feed on cocoa flowers. Eggs of *P. szentivanyi* hatch in about two weeks, larval development taking five to nine months. The pupal period is about two weeks, and sexual maturation of the adult is about eleven weeks (average times).

Several species of *Pantorhytes* seem to favour *Pipturus argenteus* as the most important local host (Szent-Ivany, 1956, 1961).

Currently control is by a multiple approach. To kill adults trichlorphon (1·7 kg in 23 litres per ha; 1·5 lb, 2 gallons, per acre) is applied using a motor-powered mistblower fitted with a special delivery restrictor. This has the advantage over more conventional higher volume delivery that droplet size being smaller and more uniform there is a higher rate of active ingredient per droplet. The infested tree is a reservoir of weevils from which, because of the long duration of the larval stage, adults will continue to emerge at an undiminished rate for some months after commencement of spraying. These adults continually augment those which have survived previous spraying (kill efficiency per application of trichlorphon is only about 70 to 80 per cent) and this composite residual population puts back some larvae into the trees. When populations have been considerably reduced by spraying at intervals of six weeks, hand collection of adults and larvae and banding trees with Ostico[R] banding grease commences.

Areas occupied by the ground-nesting ant *Anoplolepis longipes* tend to have very much smaller infestations of weevils (Baker, 1972).

In new plantings it is essential to plan for the inclusion of peripheral barrier crops (*Imperata* grass, sweet potato or possibly *Pueraria*) and the prior eradication of alternative wild host plants within the plantation area.

Ambrosia beetles

Attack by these little black cylindrical beetles, which can be identified by their small round entrance holes in the trunk or branches of trees and often in stems of seedlings, is by a very large number of species, the majority of which is attracted only to plants already either dying or dead. Thus most attack follows on severe water stress, fungal damage (stem or roots), mechanical damage and extreme senescence and is thus of a secondary nature. However, in West Africa *Xylosandrus compactus* (= *Xyleborus morstatti*) (Fig. 11.18), best known as a stem-borer of

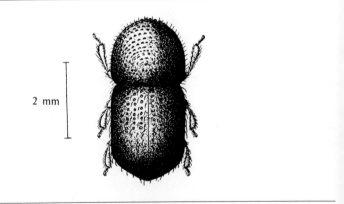

2 mm

Fig. 11.18   *Xylosandrus compactus.* Adult female.

coffee, may be a primary pest in cocoa seedlings. These are especially prone to attack when shaded and also when in weedy ground. The extent to which the beetles are successful in completing their galleries in seedling stems seems to vary regionally. For instance, in Sierra Leone and Eastern Nigeria serious damage may be done because the beetles are very successful, but in Western Nigeria and Ghana, though the incidence of attack may be high, the effect is slight because the beetles tend to be unsuccessful in excavating galleries. Maintaining a cordon sanitaire of cleared land or another crop, provided this is not coffee, between cocoa and bush or coffee areas will help in avoiding infestation. Chemical control has proved very difficult (Entwistle, 1960, 1972).

In some areas of the New World attack by *Xyleborus ferrugineus*, and to a less extent by certain other species, increases the effect of the vascular wilt fungus *Ceratocystis fimbriata*. This is brought about by beetle attack causing wounds by which the fungus can infect the tree. Furthermore, the danger of infection is increased by the large quantity of spores passed out of the beetle galleries with extruded wood dust. In other areas, notably Venezuela, beetles seem less involved in the spread of this disease. Beetle attack is primarily concentrated in the collar and lower trunk region (Iton, 1959, 1961; Saunders, 1965).

In Ecuador about 70 per cent disease control has been achieved in trials with 1 per cent BHC with a sticker (Estab[R]) applied to the tree trunk every six months. Endosulfan (same application rate and timing) gave almost equally good results (Saunders *et al.*, 1967). Combined with the plantation sanitation practices described under 'Ceratocystis wilt' (p. 147) this may well give an adequate level of control.

There is evidence that trees with Criollo characteristics are the most susceptible.

Termites or white ants

Termites are general in cocoa plantations but are usually minor pests. Attack takes two main forms. Firstly young plants, both in the nursery and in the field, may be attacked at the collar and also on the tap root and basal stem (e.g. by *Macrotermes bellicosus* in West Africa). Similar damage may be seen on chupons arising from the base of mature trees. Attack of this type is mainly a dry season phenomenon and may pass unnoticed until swift and severe wilt symptoms appear. The other form of attack concerns establishment of termite colonies within the trunk and larger branches of mature trees. Dry wood termites may attack much of the wood dying following injury of various kinds whilst invasion of the living wood by damp wood termites is generally from infestations in wounded or dead tissues. Occasionally termites attacking the collar region may spread inside the aerial framework of the tree (e.g. in Samoa), but it has generally been observed that spread occurs downwards from the point of entry. In both types of attack on living wood damage can be extreme before discovery. Both water stress and senescence seem to hasten the progress of attack by lowering the resistance of the tree.

Persistent attack in nurseries can be overcome by using potting soil containing insecticidal dusts (e.g. dieldrin at 1·2 kg of 2·5 per cent dust per m³ soil). Where large volumes of soil must be prepared a concrete mixer will be found useful. In the field seedlings are best protected by ensuring sufficient protection from desiccation, e.g. by good nurse or permanent shade cover. The treatment of damage in mature trees should be an organised part of plantation maintenance. It consists of careful pruning of dead wood, preferably with saw cuts close to the branch origin, and treatment of cut surfaces with a fungicide until callused over. Active infestations in trunks, and large troublesome nests in the soil, can be eradicated by infiltrating with 0·05 per cent aqueous dieldrin.

Snails

Minor damage by snails is widespread but the giant African Snail (*Achatina fulica*) may be very destructive of young seedlings as, for example, in Papua–New Guinea and the Caroline Islands. In Sabah it commonly feeds on young *Erythrina* shade. It may be kept out of nurseries by a low copper screen.

Vertebrates

In addition to insects much damage can be caused to cocoa by vertebrates. Elephants, wild cattle, deer, pigs, monkeys, bats, rodents, civet cats, some marsupials, woodpeckers and parrots have all been involved at one time or other. However, with increasing pressure from man numbers of the larger species especially have decreased, so that in general they now cause only local damage. This can be severe enough as, for instance, with new plantations damaged by small deer and bearded pigs in Sabah.

Primary damage to pods is experienced in most parts of the world and is usually caused by small mammals especially squirrels and rats, but sometimes monkeys. Ripe pods are usually selected for attack and a large hole is bitten through the husk. The beans are then partially or completely extracted and after the sweet surrounding pulp has been sucked off they are discarded. Green pods are less attractive but if attacked then the beans are generally eaten. Some feeble or less adaptable mammals are themselves unable to open pods but may extend existing damage and remove more of the pod contents.

Monkeys, some squirrels and in Africa the Pouched or Giant Rat (*Cricetomys gambianus*) may remove the pod intact, possibly pulling it off together with a strip of bark. Monkeys are also said to bite off the apex of the pod, whereas most rodents enter it from the side. Probably there is no means by which one can distinguish, from a damaged pod, between squirrel and rat attack.

Cocoa adjacent to forest suffers the heaviest damage, largely because it is in the forest that most species actually live. If only because of its seasonal cropping pattern cocoa forms only a part of the diet of mammalian pest species. Such species as have accepted cocoa as a permanent habitat appear to have done so mainly where it is grown beneath mixed forest shade trees and where some of the features of undisturbed forest are retained. A cleaned strip, such as a roadway, between forest and plantation may be sufficient appreciably to discourage attack, though not all species, especially if they happen to be nocturnal, object to crossing open ground (Everard, 1968; Taylor, 1961).

Pod losses may commonly be over the 5 per cent level. In Sierra Leone it is believed that during the last half century there has been an increase in the monkey population which during the past fifteen years has been accelerated. Two factors may have contributed to this; leopards which were once plentiful have been largely hunted out for their skins, and with an advance in Islam the eating of monkeys is forbidden to an increasing proportion of the people. As a result losses of 20 to 50 per cent have been attributed to joint monkey and squirrel attack.

Nursery areas may be protected from rodent damage by use of a cleared boundary strip and a fence of half-inch mesh netting, at least 3 ft high and with the basal 6 in buried in the soil.

In plantations the problem can be ameliorated by a joint system of clearing a boundary strip, including the felling of forest trees overhanging the cocoa, and the use of traps or poison. The use of diffusely applied poisons is expensive and indefensible on grounds of the general danger involved. Hence poisons are usually presented in baits laid down at baiting points. When poisons such as white arsenic (arsenous oxide or arsenic trioxide) and sodium fluoroacetate are used rats may develop bait-shyness. To avoid this, bait alone should be offered first and only when a feeding habit is established should poison be administered. A subsequent control cycle should follow the same rule but should use a different type of bait. Rats treated with sublethal doses do not generally develop bait-shyness and this explains the success of anticoagulants whose effectiveness depends on the ingestion of a series of small doses. These compounds, which are coumarone derivatives, kill by interfering with the action of vitamin K and so reduce the coagulating properties of blood with the result that trivial injury can cause a fatal haemorrhage. Rats (and pigs) for instance are killed by a dose of 1 mg per kg of body weight for five days and cats by three times the dose for the same period. One of the best known anticoagulants is Warfarin which is formulated as a 0·1 and 0·5 per cent dust for admixture with a protein-rich bait such as corn meal at a final concentration of 0·025 to 0·005 per cent, or as a 1 per cent dust for use in holes and runs. A single dose has no effect on man or domestic animals. Vitamin K is an antidote.

Monkeys have been subject to control by organised shoots in Sierra Leone, but it is doubtful if this would be effective against birds who are perhaps less easily discouraged and whose numbers may be recruited from surrounding areas.

Lizards may cause faecal contamination of the crop during sun drying and also when, in store, beans spilled from rat damaged sacks are swept up and replaced. It is as well to remember that in Africa the common Agama lizard (*Agama agama*) is a known carrier of salmonellosis.

References

This list of references includes several not mentioned in the text; they provide additional guidance on the subject.

General:

Alibert, H. (1951) 'Les insectes vivant sur les cacaoyers en Afrique Occidentale', *Mém. Inst. fr. Afr. Noire*, **15**, 174 pp.

Bondar, G. (1939) 'Insetos daninhos e parasitas do cacau na Bahia', *Bolm. téc. Inst. Cacau, Bahia, Brazil*, **5**, 112 pp.

Clayphon, J. E. (1971) 'Comparison trials of various motorised knapsack mistblowers in the Cocoa Research Institute of Ghana', *Pestic. Abstr.*, **17**, 209–25.

Conway, G. R. (1971) 'Pests of cocoa (*Theobroma cacao* L.) in Sabah and their control, with a list of the cocoa fauna', *Bull. Dep. Agric., Sabah.* 125 pp.

Dinther, J. B. M. van (1960) 'Insect pests of cultivated plants in Surinam', *Bull. Landbouwproefst. Sur.*, **76**, 159 pp.

Downing, S. F. (1967) 'The development of pesticides. A chemical manufacturer's viewpoint', *Cocoa Growers' Bull.*, **9**, 19–24.

Entwistle, P. F. (1972) *Pests of Cocoa*, Longman.

Gerard, B. M. (1969) 'Tests for flavour in chocolate and residues in cacao beans resulting from the use of insecticides on cacao trees', *Mem. 2nd Internat. Cacao Res. Conf., Bahia, Brazil, 1967*, pp. 519–25.

Hancock, B. L. (1968) 'The development of pesticides. The role of the chocolate manufacturer', *Cocoa Growers' Bull.*, **10**, 17–21.

Higgins, A. E. H. (1964) 'The selection of spraying equipment, Part 1', *Cocoa Growers' Bull.*, **3**, 21–6.

Higgins, A. E. H. (1965) 'The selection of spraying equipment, Part 2', *Cocoa Growers' Bull.*, **4**, 20–5.

Johnson, C. G. (1962) 'The ecological approach to cocoa disease and health', in J. B. Wills, ed., *Agriculture and Land Use in Ghana*, pp. 348–52.

Kalshoven, L. G. E. (1950–51) *De plagen van de cultuurgewassen in Indonesie*, 2 vols, N.V. Uitg., The Hague, W. van Hoeve.

Leston, D. (1970) 'Entomology of the cocoa farm', *Ann. Rev. Ent.*, **15**, 273–94.
Martin, H. (1968) *Pesticide Manual*, Brit. Crop Protection Council.
Miller, N. C. E. (1941) 'Insects associated with cocoa in Malaya', *Bull. ent. Res.*, **32**, 1–16.
Newhall, A. G. (1966) 'When does it pay to spray cocoa?' *Cacao*, **11**, 10–12; also in *Cocoa Growers' Bull.*, 1968, **10**, 22–6.
Silva, P. (1944) 'Insect pests of cacao in the State of Bahia', *Trop. Agric. Trin.*, **21**, 8–14.
Smee, L. (1963) 'Insect pests of *Theobroma cacao* in the Territory of Papua and New Guinea: their habits and control', *Papua New Guinea agric. J.*, **16**, 1–19.
Szent-Ivany, J. J. H. (1961) 'Insect pests of *Theobroma cacao* in the Territory of Papua and New Guinea', *Papua New Guinea agric. J.*, **13**, 127–47.
Tinsley, T. W. (1964) 'The ecological approach to pest and disease problems of cacao in West Africa', *J. R. Soc. Arts*, April 1964, pp. 353–71.

Mirids (Capsids):
Collingwood, C. A. (1971) *Cocoa Capsids in West Africa. Report of International Capsid Research Team 1965–71*, London, Internat. Office Cocoa and Chocolate.
Entwistle, P. F. and Youdeowei, A. (1964) 'A preliminary world review of cacao mirids', *Proc. Conf. Mirids and Other Pests of Cacao, Ibadan, Nigeria, 1964*, pp. 71–9.
Gibbs, D. G., Pickett, A. D. and Leston, D. (1968) 'Seasonal population changes in cocoa capsids (Hemiptera, Miridae) in Ghana', *Bull. ent. Res.*, **58**, 279–93.
Lavabre, E. M., Decelle, J. and Debord, F. (1963) 'Etude de l'évolution régionale et saisonière des populations des mirides (capsides) en Côte d'Ivoire', *Café Cacao Thé*, **7**, 267–89.
Meer Mohr, J. C. van de (1927) 'Au sujet du rôle de certaines fourmis dans les plantations coloniales', *Bull. agric. Congo Belge*, **28**, 97–106.
Mire, P. Bruneau de (1965) 'Comparaison entre deux modes de traitment anti-mirides du cacaoyer: la thermonebulisation et l'atomisation', *Conf. Internat. Recherches Agron. Cacaoyeres, Abidjan, Côte d'Ivoire, 1965*, pp. 154–9.
Mire, P. Bruneau de (1969) 'Une fourmi utilisée au Cameroun dans la lutte contre les mirides du cacaoyer, *Wasmannia auropuncta* Roger', *Café Cacao Thé*, **13**, 209–12.
Nicol, J. and Taylor, D. J. (1954) 'Capsids and capsid control in the Belgian Congo with special reference to Lukolela Plantations', *Tech. Bull. W. Afr. Cocoa Res. Inst.*, **2**, 10 pp.
Stapley, J. H. and Hammond, P. S. (1959) 'Large scale trials with insecticides against capsids on cacao in Ghana', *Emp. J. exp. Agric.*, **27**, 343–53.

Shield Bugs:
Brown, E. S. (1958) 'Injury to cacao by *Amblypelta* Stål (Hemiptera, Coreidae) with a summary of food-plants of species of the genus', *Bull. ent. Res.*, **49**, 543–54.
Callan, E. McC. (1944) 'Cacao stink bugs (Hem., Pentatomidae) in Trinidad, B.W.I.', *Revta Ent., Rio de J.*, **15**, 321–4.
Franco, T. H. de (1958) 'Transmission de la moniliasis del cacao por el *Mecistorhinus tripterus* F.', *Proc. 7th Inter-Amer. Cacao Conf., Palmira, Colombia, 1958*, pp. 130–6.
Lodos, N. (1967a) 'Studies on *Bathycoelia thalassina* (H.-S.) (Hemiptera, Pentatomidae), the cause of premature ripening of cocoa pods in Ghana', *Bull. ent. Res.*, **57**, 289–99.
Lodos, N. (1967b) 'Contribution to the biology of and damage caused by the cocoa coreid, *Pseudotheraptus devastans* Dist. (Hemiptera-Coreidae)', *Ghana J. Sci.*, **7**, 87–102.
Marchart, H. and Lodos, N. (1969) 'The biology and insecticidal control of the cocoa pod pentatomid *Bathycoelia thalassina* (Herrich-Schaeffer) (Hemiptera, Pentatomidae)', *Ghana J. Agric. Sci.*, **2**, 31–7.
Sepulveda, R. L. (1955) 'Biologia del *Mecistorhinus tripterus* F. (Hem. Pentatomidae) y su posible influencia en la transmisión de la Moniliasis del cacao', *Cacao, Colomb.*, **4**, 15–42.

196   *Insects and Cocoa*

Leaf Hoppers, Psyllids and Aphids:

Cotterell, G. S. (1943) 'Entomology', *Rep. Centr. Cocoa Res. Stn., Tafo, Ghana, 1938–42*, pp. 46–55.

Fernando, H. E. (1959) 'Studies on *Empoasca devastans* Dist. (Fam. Jassidae, ord. Hemiptera), a new pest of cacao causing defoliation and its control', *Trop. Agric. Mag. Ceylon Agric. Soc.,* **115** (2), 121–44.

Kirkpatrick, T. W. (1955) 'Notes on minor insect pests of cacao in Trinidad, Part 3, Aphididae', *Rep. Cacao Res., Trinidad 1954*, pp. 56–7.

Lodos, N. (1969) 'Minor pests and other insects associated with *Theobroma cacao* L. in Ghana', *Ghana J. Agric. Sci.,* **2**, 61–72.

Salas, A. and Hansen, A. J. (1963) 'A toxicogenic leaf-hopper (Homoptera: Cicadellidae) observed on cacao (*Theobroma cacao* L.)', *Cacao, Turrialba, Costa Rica,* **8** (1), 6–12.

Mealybugs and Scale Insects:

Entwistle, P. F. (1973) 'Coccoids', in A. J. Gibbs, ed., *Viruses and Invertebrates,* Amsterdam, North-Holland Publishing Company.

Strickland, A. H. (1947) 'Coccids attacking cacao (*Theobroma cacao* L.) in West Africa, with descriptions of five new species', *Bull. ent. Res.,* **38**, 497–523.

Strickland, A. H. (1951) 'The entomology of swollen shoot of cacao, I. The insect species involved, with notes on their biology', *Bull. ent. Res.,* **41**, 725–48.

Strickland, A. H. (1951) 'The entomology of swollen shoot of cacao, II. The bionomics and ecology of the species involved', *Bull. ent. Res.,* **42**, 65–103.

Thrips:

Cotterell, G. S. (1928) 'The Red Banded Cacao Thrips, *Heliothrips rubrocinctus*, Giard', *Yb. Dep. Agric. Gold Cst, 1927*, pp. 94–9.

Cotterell, G. S. (1930) 'Notes on the occurrence of *Heliothrips rubrocinctus*, Giard, in San Thomé, and possibility of control by biological means', *Yb. Dep. Agric. Gold Cst., 1929*, pp. 130–3.

Fennah, R. G. (1947) 'The insect pests of food-crops in the Lesser Antilles', Depts. of Agric. Windward and Leeward Islands 1947.

Fennah, R. G. (1955) 'The epidemiology of Cacao-thrips on cacao in Trinidad', *Rep. Cacao Res., Trinidad 1954*, pp. 7–26.

Fennah, R. G. (1963) 'Nutritional factors associated with seasonal population increase of cacao thrips, *Selenothrips rubrocinctus* (Giard) (Thysanoptera), on Cashew, *Anacardium occidentale*', *Bull. ent. Res.,* **53**, 681–713.

Fennah, R. G. (1965) 'The influence of environmental stress on the cacao tree in predetermining the feeding sites of cacao thrips, *Selenothrips rubrocinctus* (Giard), on leaves and pods', *Bull. ent. Res.,* **56**, 333–49.

Reyne, A. (1921) 'De cacaothrips (*Heliothrips rubrocinctus*, Giard)', *Bull. Dep. Lanbouw. Suriname,* **44**, 214.

Caterpillars:

Catley, A. (1963a) '*Tiracola plagiata* Walk. (Lepidoptera: Noctuidae). A serious pest of cacao in Papua', *Papua New Guinea agric. J.,* **15**, 15–22.

Catley, A. (1963b) 'Observations on the biology and control of the armyworm *Tiracola plagiata* Walk. (Lepidoptera: Noctuidae)', *Papua New Guinea agric. J.,* **15**, 105–9.

Dun, G. S. (1967) 'Cacao defoliating caterpillars in Papua and New Guinea', *Papua New Guinea agric. J.,* **19**, 67–71.

Entwistle, P. F. (1969) 'The biology of *Earias biplaga* Wlk. (Lep., Noctuidae) on *Theobroma cacao* L. in Western Nigeria', *Bull ent. Res.,* **58**, 521–35.

Kalshoven, L. G. E. (1919) 'De roode takboorder, *Zeuzera coffeae* Nietener in boschcultuuren. De roode stamboorer, *Zeuzera postexcisa* Hamps', *Meded. Proefst. Boschwezen,* **4**, 57–65.

Kalshoven, L. G. E. (1922) 'Schade door den "Ringboorder" *Phassus* (?) *damor* Moore, aan *Wildboutculturen*', *Meded. Proefst. Boschwezen*, **4**, 75–81.
Kalshoven, L. G. E. (1940) 'Observations on the Red Branch-Borer, *Zeuzera coffeae* Nietn', *Ent. Meded. Ned-Indie*, **6**, 50–4.
Laoh, J. P. (1954) 'The control of the pod borer *Acrocercops cramerella* in cacao' (in Dutch), *Bergcultures*, **23** (8) and **23** (24); also as a Shell Co. *Agric. Bull.*
Nguyen-Ban, J. (1971) 'Essais en insectarium d'insecticides systemiques contre les chenilles d'*Earias biplaga* (Wlk.) et les cochenilles blanches; *Pseudococcus* sp.', *Proc. 3rd Internat. Cocoa Res. Conf., Accra, Ghana, 1969*.
Roepke, W. (1912) 'Over den huidingen stand van het vraagstuk van het rampassen als bestrijdinge middel tegen de cacao mot op Java', *Meded. Proefst. Midden-Java*, **8**, 1–21.

Leaf-cutting Ants:
Cherrett, J. M. (1968) 'The foraging behaviour of *Atta cephalotes* L. (Hymenoptera, Formicidae). I. Foraging pattern and plant species attacked in tropical rain forest', *J. Anim. Ecol.*, **37**, 387–403.
Cherrett, J. M. (1969) 'Baits for the control of leaf-cutting ants, I. Formulation', *Trop. Agric.*, **46**, 81–90.
Cherrett, J. M. and Merrett, M. R. (1969) 'Baits for the control of leaf-cutting ants, II. Waterproofing and general broadcasting', *Trop. Agric.*, **46**, 221–31.
Cherrett, J. M. and Sims, B. G. (1968) 'Some costings for leaf-cutting ant damage in Trinidad', *J. Agric. Soc. Trin.*, **68**, 313–22.
Cherrett, J. M. and Sims, B. G. (1969) 'Baits for the control of leaf-cutting ants, II. Toxicity evaluation', *Trop. Agric.*, **46**, 211–19.

Enxerto, Cacarema and Balata Ants:
Silva, P. (1957) 'A "formiga de enxerto"', *Divulg. Inst. Cacau Bahia*, **1**, 15 pp.
Silva, P. and Bastos, G. A. C. (1965) 'Polyvilhadeira-injetora PG para combater formiga-de-enxerto', *Cacau Atual*, **2**, 20–3.

Rose Beetle and Other Chafers:
Cotterell, G. S. (1930) 'Report on the occurrence of *Sahlbergella* spp. and other insect pests of cacao in Fernando Po, San Thomé and the Belgian Congo', *Yb. Dep. Agric. Gold Cst., 1929*, pp. 112–33.
Lever, R. J. A. W. (1953) 'Cockchafer pests of cacao and other crops', *Malay agric. J.*, **36**, 89–113.
Nosti, J. (1953) *Cacao cafe té*, Madrid, Salvat Editores, 687 pp.
O'Connor, B. A. (1959) 'Insect pests of cocoa', *Agric. J. Fiji*, **29**, 92–4.

Cocoa Beetle:
Fennah, R. G. (1948) 'Studies on measures for control of cacao-beetle', *Food Crop Pest Investigations, Windward and Leeward Islands, Final Report* (stencilled), 10 pp.
Fennah, R. G. (1954) 'Studies on cacao beetle (*Steirastoma breve* Sulz.)', *Rep. Cacao Res., Trinidad 1953*, pp. 73–9.
Guppy, P. L. (1911) 'The life history and control of the cacao beetle', *Circ. Bd. Agric., Trin.*, **1**, 1–34.
Urich, F. W. (1925) 'The cacao beetle', *Bull. Dep. Agric., Trin.*, **21** (1), 36–9.

Glenea Longhorn Borers:
Dun, G. S. (1951) 'Pests of cacao in the Territory of Papua and New Guinea', in D. H. Urquhart and R. E. P. Dwyer, eds., *Prospects for Extending the Growing of Cacao in Papua and New Guinea*, Bournville, Cadbury Brothers, pp. 32–5.
Schreurs, J. (1965) 'Investigations on the biology and control of *Glenea lefebueri*, a noxious longicorn beetle of cacao in West New Guinea', *Papua New Guinea agric. J.*, **17**, 129–35.

Weevil Borers:

Baker, G. (1972) 'The role of *Anoplolepis longipes* Jordan (Hymenoptera: Formicidae) in the entomology of cacao in the Northern District of Papua–New Guinea', *Abstracts 14th Internat. Congr. Entomology*, p. 327.

Gressitt, J. L. (1966) 'The weevil genus *Pantorhytes* (Coleoptera), involving cacao pests and epizoic symbiosis with cryptogamic plants and microfauna', *Pacif. Insects*, **8**, 915–65.

Szent-Ivany, J. J. H. (1956) 'Two new stem borers of cacao in New Guinea', *FAO Plant Prot. Bull.*, **4**, 177–8.

Szent-Ivany, J. J. H. (1961) 'Insect pests of *Theobroma cacao* in the Territory of Papua and New Guinea', *Papua New Guinea Agric. J.*, **13**, 127–47.

Ambrosia Beetles:

Entwistle, P. F. (1960) 'A review of the problem of shot hole borer (Coleoptera, Scolytidae and Platypodidae) attack in cocoa in West Africa', *Proc. 8th Inter-Amer. Cacao Conf., Trinidad*, pp. 208–23.

Iton, E. P. (1956) 'Studies on wilt disease of cacao at River Estate', *Rep. Cacao Res., 1957–58, Trinidad*, pp. 55–64.

Iton, E. P. (1961) '*Ceratostomella* wilt in cacao in Trinidad', *Proc. 8th Inter-Amer. Cacao Conf., Trinidad*, pp. 201–4.

Saunders, J. L. (1965) 'The *Xyleborus-Ceratocystis* complex of cacao', *Cacao*, Turrialba, Costa Rica, **10** (2), 7–13.

Saunders, J. L., Knoke, J. K. and Norris, D. M. (1967) 'Endosulfan and lindane residues on the trunk bark of *Theobroma cacao* for the control of *Xyleborus ferrugineus*', *J. Econ. Ent.*, **60**, 79–82.

Damage by Vertebrates:

Everard, C. O. R. (1968) *A Report on the Rodent and other vertebrate pests of cocoa in Western Nigeria*, Ibadan, Nigeria, Res. Divis., Min. Agric.

Taylor, K. D. (1961) 'An investigation of damage to West African cocoa by vertebrate pests' (cyclostyled report).

Table 11.1   *Insecticides referred to in this chapter*

	LD_{50} to white rats*
Organo-chlorine (= chlorinated hydrocarbons)	
aldrin	50
gamma-BHC	125
chlordane	283
dieldrin	100
endosulfan	40–50
endrin	10–12
heptachlor	100
DDT	113
Organo-phosphates	
fenitrothion	250
menazon	1,950
trichlorphon	625
Carbamates	
carbaryl	560
Orthobux (Ortho 5353)	87
propoxur (arprocarb)	100

Table 11.1 *Insecticides referred to in this chapter*—continued

	LD_{50} *to white rats**
Miscellaneous	
'coal tar derivatives' (presumably tar oils)	apt to cause dermatitis, especially in sunlight
lead arsenate	10–15
methyl bromide	respirators to be worn at concentrations over 17 ppm
mirex	306
sodium fluoroacetate	0·22
Warfarin	1 per day for 5 days
white arsenic	20–300

*The dose of toxicant which causes 50 per cent mortality. It is measured in terms of milligrams per kilogram of body weight.

The following figures are a guide as to the level of toxicity:

Less than 10	Extremely toxic
11–50	Highly toxic
51–500	Moderately toxic
501–5,000	Slightly toxic

Chapter 12

Harvesting and Preparation

Development of crop

In common with most other tropical crops, the cocoa harvest is not confined to one short period but is spread over several months. There are peak harvest periods, one or two per year, and in many countries there is some cocoa to be harvested at all times of the year.

The interval between fertilising the flower and harvesting the ripe pod is about five months. Early measurements showed the period of development to be sixteen to twenty-one weeks in Nigeria (Waters and Hunter, 1929), five to six months in Grenada (Cooper, 1940), and an average of six months in Papua–New Guinea (Bridgland, 1953). There is considerable variation in the maturation period and there is evidence that this is correlated with maximum temperature, pods growing more slowly in the cooler months. In Bahia pods of the 'temperao' crop develop during warmer months and mature in 140 to 175 days, whereas pods of the 'safra' crop take 167 to 205 days to mature. The following formula has been evolved to calculate the maturation period for the Catongo variety in Bahia (Alvim *et al.,* in press):

$$N = \frac{2,500}{T-9}$$

where N = number of days to maturity
T = daily mean temperature

In countries with a pronounced wet and dry season the main harvest will occur five to six months after the start of the wet season and correlations between rainfall and production five months later have been established (Bridgland, 1953). This results from a burst of flowering at the beginning of the wet season and studies of climate and flowering showed that flowering was dependent on temperature and rainfall (Alvim, 1966). It has been assumed that flowering occurred

200

when the monthly mean temperature exceeds 23°C and the monthly rainfall is greater than the potential evapotranspiration. On this basis most flowering patterns can be explained, but there are anomalies, in particular the lack of flowering in West Africa in the hot wet months of October and November. It has recently been shown that this is due to the crop on the tree suppressing flowering (Alvim *et al.,* in press).

In addition to the three factors already mentioned, crop pattern is affected by the variety of cocoa. In Nigeria Amelonado cocoa has a sharply peaked crop 28 per cent and 19 per cent of the crop being harvested in November and December respectively; on the other hand F_2 Amazon has a more widely spread crop, producing 20 per cent in January and 16 per cent in May (Toxopeus, 1964).

The crop pattern for Ghana is shown in Fig. 12.1. The data for Ghana

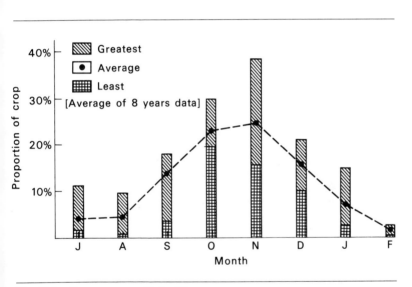

Fig. 12.1   Crop pattern at Tafo, Ghana.

show that on average 25 per cent of the crop is harvested in the peak month, November, which is about six months after the wet season begins. In West Cameroon the pattern is similar but data for several years show that the proportion of the crop gathered in the peak month can vary between 20 and 45 per cent. In Malaya there is no true dry season; the peak of cropping is less pronounced, with 20 to 25 per cent of the crop in the peak month, which falls between November and March. The less pronounced peak has advantages in spreading the task of harvesting and in reducing the capacity of the fermentary and dryers which must be sufficient to handle the largest harvest.

Harvesting

Harvesting involves removing ripe pods from the trees and opening them to extract the wet beans. On ripening the pods change colour, green pods becoming orange-yellow and red pods turning orange, particularly in furrows. The colour changes with red pods are less obvious and in a mixed population of trees there may be some initial difficulty in identifying ripe pods. However, the time for harvest is not critical; pods which are not fully ripe will ferment as well as ripe pods; and ripe pods can be left on the tree for two to three weeks. If left any longer the beans may germinate inside the pods. This occurs occasionally, usually in a severe dry season. There is, therefore, a considerable period of time, three to four weeks, during which the pod is fit for harvest.

The frequency of harvesting is not determined solely by this factor. The weight of cocoa gathered at one harvest affects the harvesting cycle; a certain minimum of weight, about 100 kg (2 cwt) wet beans, is needed for satisfactory fermentation and on small farms harvesting may have to be carried out at intervals longer than one month in order to collect this amount. On the small farms of West Africa the usual practice is to harvest only three or four times in the course of the main crop season. Such infrequent harvesting may well be detrimental to quality and to yield as some pods will have developed pod rots or have been damaged by rodents. On larger farms and estates harvesting is done at intervals no longer than three to four weeks and weekly harvesting is the rule where pod diseases or rodents are liable to cause appreciable losses.

Yield may be influenced by frequency of harvesting but the evidence is conflicting. A comparison of weekly and monthly harvesting at Tafo showed a clear advantage in number of pods per tree over two seasons but later experiments failed to establish this (Wickens, 1955; McKelvie, 1958).

Pods are removed from the tree by various forms of knife; a short-handled cutlass or machete is often used for the pods within reach and special harvesting knives on long poles for pods in the branches. It is advisable to exercise care to avoid damage to the flower cushion when pods are removed as wounds will provide entry for fungi.

After 'plucking' the pods are opened, but practice varies considerably. The commonest practice is to gather the pods to one or more convenient places in the field and to open them when the round of harvesting has been completed. This may involve leaving the pods unopened for a few days; this is not detrimental unless pod diseases are prevalent, in which case some loss of pods may result. It has been found that a delay in opening the pods results in a more rapid temperature rise in fermentation and has a significant effect on purple bean percentage.

Table 12.1   *Effect of storage of pods on purple bean percentage*

	No. of days storage			
	1	*2*	*3*	*4*
Purple beans %	82·0	44·3	39·6	52·8

SOURCE: MacLean and Wickens (1951).

In another trial pods were kept for a week before opening and no difference was found in the temperature curve during fermentation nor in the flavour of chocolate made from the beans (Howat *et al.*, 1957a). There may, therefore, be some advantage in delaying pod opening and in Papua–New Guinea an interval of three to four days is recommended. In other countries such a delay may lead to larceny so that the practice cannot be employed.

As an alternative to opening in the field, the pods may be transported to the fermentary and opened there. This alternative has the advantage of greater control over the process as it can be done under cover, but it involves the transport of four times as much weight in comparison with wet beans and some means of disposing of pod husks must be found.

A further alternative is to open the pods as soon as they are plucked from the trees. The considerable advantage of this method is the saving in transport and the distribution of pod husks. There is virtually no transport of pods as such, only the wet beans, and this should improve productivity. The pod husks are distributed throughout the fields by this method thus returning the nutrients in the pod husks which are rich in potash. It is, however, more difficult to supervise with regard to quality and ripeness of pods.

Pods are usually opened by cutting them diagonally with a knife which is then twisted, thereby breaking off a portion of the pod. The beans are then removed from the pod with the fingers and this is often done by a second person. While this is the commonest method of pod opening, the use of a knife can result in some damage to the beans; this may rise as high as 5 per cent of beans with a cut testa, and such splits provide an entrance for moulds and stored product pests (Wadsworth, 1953). To avoid this, pods can be opened by cracking them on a stone or with a wooden billett or on to a triangular strip of wood fixed to a table.

The beans are joined to a placenta in the pod. If the placenta is removed from the pod it should be separated from the wet beans, preferably before fermentation. Its presence will not interfere with fermentation but it may prove more difficult to remove at a later stage.

The process of pod opening is one that would appear suitable for mechanisation and no doubt this will be achieved in time, but no simple and effective machine has been developed. Some attempts at developing a pod-opening machine have been reported (Jimenez, 1967; Jabogun, 1965; Wood, 1968) but so far they do not appear to have been successful, which may be due to the difficulty of separating wet beans from pieces of broken husk.

Where the tasks of harvesting and pod opening are separate it has been found that a man can harvest 1,500 pods a day, and similarly a man can open 1,500 pods a day. The two jobs will require thirty-three man-days per ton and in addition there will be the cost of transporting pods or beans to the fermentary. Where pods are opened immediately after plucking the labour requirements are slightly less; one man can harvest and open 900 pods a day, giving a labour requirement of thirty man-days per ton.

These figures are based on yields of 800 to 900 kg dry cocoa per hectare (700 to 800 lb per acre) and are an average for the whole crop. The number of pods a man can harvest in a day will vary with the level of cropping; it will be higher during the peak season and lower at other times when the harvesters have to walk further to collect their harvest.

Fermentation

At one time it was assumed that cocoa beans were fermented in order to get rid of the pulp around the beans. Though this is one result of fermentation it is not the main purpose, which is to produce cocoa beans which on manufacture will produce good chocolate. Chocolate flavour is developed by the two processes of fermentation by the grower and roasting by the manufacturer and good flavour cannot be obtained by one of these processes alone.

In brief, fermentation involves keeping a mass of cocoa beans well insulated so that heat is retained, while at the same time air is allowed to pass through the mass. The process lasts up to seven days and is followed immediately by drying. The chemical processes involved are complex and incompletely understood; the term fermentation, which normally means the conversion of sugar to alcohol and from alcohol to acetic acid, correctly describes the changes that take place in the pulp, but there are many other quite different changes which take place inside the bean.

In practice the methods of fermentation vary enormously from country to country, or from one grower to another. Some methods have been evolved for special circumstances and cannot be used successfully under others; but the wide variation that remains indicates the variety of factors involved and the latitude that applies to some of them. While the process should not be treated casually, neither should it be taken too seriously, particularly at the early stages of production when the quantities available for fermentation are minimal. No method of fermentation will follow the same course on every occasion and some of the difficulties which arise need not be treated by varying the techniques but will simply disappear. The process of fermentation is, after all, dependent on the activities of a succession of micro-organisms and it is perhaps more remarkable that the process should generally succeed than that there should be occasional failures.

Changes occurring during fermentation

When the beans are removed from the pod they are covered with mucilage or pulp. The proportion of the whole beans formed by the pulp can be gauged from the losses during fermentation which consist largely of the water in the pulp and the products of changes of some constituents of the pulp. These losses amount to 15 to 20 per cent of the weight of the wet beans and will vary according to season and the type of cocoa.

The pulp consists of 80 per cent water, 10 to 15 per cent glucose and fructose, up to 0·5 per cent non-volatile acid, largely citric, and small amounts of sucrose, starch, volatile acids and salts. The pulp is sterile initially, but the presence of sugars and the high acidity (pH 3·5) provide excellent conditions for the development of micro-organisms. A wide range of micro-organisms infect the mass of beans through the activity of fruit flies and contamination from the fermentary. In the first stages yeasts proliferate and they convert sugars in the pulp to alcohol. The cells of the pulp start to break down soon after the fermentation process begins either through an enzyme change or by simple mechanical pressure. The liquid portion of the pulp runs off as 'sweatings' and the flow of sweatings is normally completed during the first twenty-four to thirty-six hours of fermentation. The activity of the yeasts leads to a large amount of carbon dioxide being evolved and at this stage of fermentation relatively anaerobic conditions prevail. These conditions allow the development of lactic acid bacteria which assist in the breakdown of sugars.

When the sweatings have run off the conditions become more aerobic and the acidity is reduced by the removal of citric acid. The presence of oxygen allows acetic acid bacteria to take over from yeasts and convert alcohol to acetic acid; this is assisted if the fermenting mass is turned. The reactions in which the micro-organisms are involved produce much heat which raises the temperature of the mass of beans (Kenten and Powell, 1960).

The temperature rises steadily during the first two days reaching 40 to 45°C at the end of that time. After the first mixing the temperature will normally rise more rapidly to 48 to 50°C. Thereafter the temperature falls slowly, but will rise again after the beans are turned. After six days the temperature of the mass of beans will be 45 to 48°C. The changes accompanying overfermentation can increase temperature to 50 to 52°C.

There is some variation in temperature within the fermenting mass and the figures given refer to the bulk of the beans. In a box the temperature is lower around the sides and close to the bottom of the box. In heaps there is greater variation as uniform aeration is impossible and Rohan (1958b) has shown how temperature differences increase with the size of the heap.

Temperature curves of box and heap fermentations are given in Figs. 12.2 and 12.3.

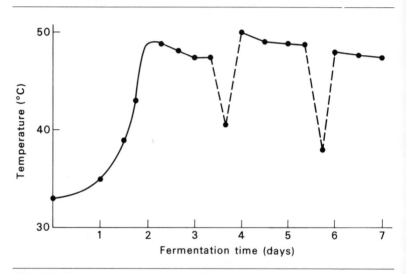

Fig. 12.2   Temperature curve of a box fermentation.

The low pH of fresh pulp has been mentioned; at the same stage the pH of the cotyledon is 6·6. As the testa is permeable to acetic acid, it passes into the cotyledon, kills the bean and lowers the pH to 4·8 by the third day. Thereafter the pH rises gradually during the remaining period of fermentation and drying and is usually 5·5 in the dried bean. The changes in pH during a box fermentation are shown in Fig. 12.4.

The death of the bean results in a breakdown of the internal cell structure which allows various enzyme reactions to take place converting some of the polyphenolic compounds. The various chemical changes in the cotyledons are vital to the development of chocolate flavour.

Numerous chemical changes take place in the cotyledons; some of the more important concern the polyphenols, of which there are several and the changes they undergo are reflected in both flavour and colour of the bean. Amongst the polyphenols are two anthocyanins which give

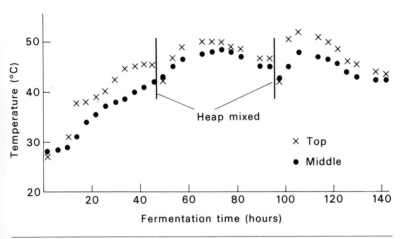

Fig. 12.3   Temperature curve for a heap fermentation.

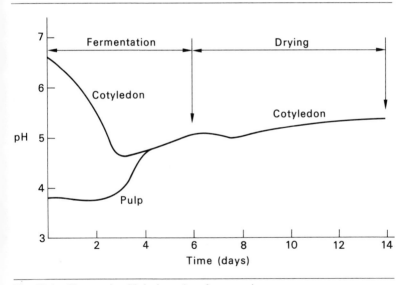

Fig. 12.4   Changes in pH during a box fermentation.

the purple colour to unfermented Forastero beans but are absent from Criollo beans. The anthocyanins are hydrolysed to give cyanidin and two reducing sugars; other polyphenols undergo chemical changes or diffuse through the testa, both of which processes help to reduce bitterness and astringency. The proteins in the cotyledon undergo hydrolysis, giving rise to amino acids, and conversion to insoluble forms by reaction with the polyphenols. While these changes have some influence on the chocolate flavour produced after curing and roasting, it is clear that chocolate flavour is dependent on many compounds and the balance between them.

There has been much discussion on the extent to which the changes are dependent on aerobic or anaerobic conditions, i.e. the presence or absence of oxygen. It has been said that fermentation in anaerobic and drying an aerobic stage, and conclusions drawn as to the need for mixing and aeration of the mass of fermenting beans. A distinction has to be drawn between conditions within the mass of beans and those inside the bean itself. In the mass of beans conditions rapidly become anaerobic as oxygen is used and carbon dioxide is produced. This takes place in the first day or two when the beans settle into a relatively impermeable mass. After the first mixing air can pass through the beans and is necessary for the oxidation of alcohol to acetic acid. Inside the bean conditions are anaerobic initially and some of the desirable changes to the polyphenols would be prevented if oxidation changes were possible in the early stages. The importance of anaerobic conditions inside the beans has been questioned (Powell, 1959) but the important point is that aerobic conditions within the mass of beans is necessary and is achieved by the use of properly designed boxes and by turning the beans.

Traditional methods of fermentation

The variety of methods of fermentation is legion but basically there are two methods which have been in common use for many years; the box and the heap. The box method is used in the West Indies and in many countries in Latin America; while it is not unknown in West Africa, the simple heap method is almost universally used there.

Fermentation in boxes involves the use of strong wooden boxes holding up to 1·5 tons of wet beans. There is provision for drainage of sweatings which also allows air to enter and pass up through the beans which are covered with banana leaves or sacking. The beans are turned either every day or every other day for six to eight days.

In the heap method beans are laid on plantain or banana leaves and covered with more leaves when the heap is complete; the leaves are normally held in place by pieces of wood. Fermentation lasts four to six days, during which time the heap may be turned once or twice.

Cocoa is also fermented in baskets lined and covered with leaves, a method used in parts of Nigeria and Ghana. Holes or shallow depressions in the ground have been used in West Africa and may occasionally be used today but the method is not a good one as the sweatings cannot drain away easily and the mass of beans is not aerated properly.

In Ecuador the traditional method of fermentation is to heap the beans on the drying floor, spreading them during the day to dry and heaping again at night. This method gave a good product with the old Nacional cocoa, but as this type has gradually given way to higher yielding Trinitarios, to which the method is not suited, the product has tended to decline in quality.

In the Dominican Republic fermentation is dispensed with, the beans being placed on the drying platform after the pods are opened. When the crop is heavy the layer of beans is thicker and some of the changes associated with fermentation occur, but in general the product is of poor quality and contains a high proportion of unfermented beans.

In Brazil box fermentation is employed but little attention is paid to aeration and insulation of the mass of beans. The boxes are not always provided with holes and the beans remain uncovered. This may account for the greater variability in Bahia cocoa and its more bitter flavour.

Enough has been said to indicate the variations in fermentation methods and it is beyond the scope of this book to go into further details of the methods used in individual countries. More detailed descriptions have been given by Forsyth and Quesnel (1957).

Factors affecting fermentation

The various methods of fermentation are themselves capable of great differences in detail and this reflects the large numbers of factors influencing the process.

Ripeness of pods   Where the harvesting rounds are done at intervals of three weeks or less, the pods should be at a fairly uniform state of ripeness but where the intervals are longer under- and over-ripe pods may be harvested.

Some trials of the effect of ripeness on fermentation have been reported but the degree of ripeness is not easy to define closely. In a trial conducted in Trinidad, Knapp (1926) found that 'wholly unripe' pods did not ferment normally, the temperature remaining at 35°C, after an initial rise to 40°C. The losses during fermentation and drying were far higher than normal so that the yield of dry beans was no more than 21 per cent of the wet weight. The bean size was also smaller at 1·05 gm compared with 1·34 gm for the overripe beans.

This indicates that the unripe pods used in this trial were not fully developed and presumably the pulp was deficient in sugar. MacLean and Wickens (1951) initiated a similar trial in Ghana, with somewhat similar results. Table 12.2 shows that the recovery or ratio of dry to wet weight was lower for unripe pods and the proportion of purple beans was higher than normal in unripe beans.

The degree of ripeness was not defined precisely in these experiments but it seems that the pods were very far from ripe, at a stage when the

Table 12.2 *Effect of ripeness of pods*

Factor	Experiment number	Degree of ripeness		
		Under-ripe	Ripe	Over-ripe
Purple bean %	1	85·2	81·3	65·6
	2	70·2	53·7	28·1
Recovery %	1	39·9	44·1	45·2
	2	39·5	43·1	44·2

SOURCE: MacLean and Wickens (1951).

pulp is still firm and the beans have not separated from the pod wall. In another trial, Howat *et al.* (1957a), fermented beans from Amelonado pods which were greenish-yellow and found no differences from normal in the fermentation nor in the dried beans.

It has already been mentioned that over-ripe pods may contain germinated beans but, from the point of view of quality, there is no other objection to them. The figures in Table 12.2 show that over-ripe pods give a lower percentage of purple beans and slightly higher recovery.

Pod diseases   Most pod diseases lead to complete loss of the beans they contain, and even when the beans are not destroyed it is undesirable to use the beans in a fermentation. In the case of black pod the beans may not always be lost as the fungus attacks the pod husk initially and if a ripe or nearly ripe pod is attacked, the beans can be saved by regular harvesting. If, however, the beans are attacked, it leads to a rise in free fatty acid and chocolate made from such beans will not have a normal chocolate flavour (MacLean, 1953).

Type of cocoa   There is a basic difference between Criollo and Forastero types in the way they are fermented. Criollo cocoa is fermented for a relatively short period of two to three days while Forastero is fermented for five to seven days, occasionally longer.

As a result of this difference mixed fermentations of the two types should be avoided. This can be arranged in places where Criollo and Forastero trees are grown separately, but the commoner situation is for pods from hybrid trees to contain both white and purple beans which are impossible to segregate.

Quantity of cocoa   The heat generated during fermentation is retained by insulation but this becomes more difficult to achieve with small quantities of beans as their surface area is great in relation to their mass. There is therefore a minimum quantity of beans which will ferment satisfactorily. Various opinions have been given as to this minimum quantity. Rohan (1958c) found that heaps containing 70 kg (154 lb) wet beans can be properly fermented but both higher and lower figures have been put forward. As a rough guide the weight of wet cocoa should not be less than 90 kg (200 lb) when the traditional box or heap methods are used. Small quantities are liable to be more affected by changes in outside temperature and require better insulation.

The maximum quantity that can be fermented will depend on the method employed. In box fermentation there is a depth above which aeration is reduced by the pressure of the beans; this maximum depth is about 0·75 m. Apart from being inadequately fermented, a greater depth of beans would be more difficult to mix and handle.

Duration   An enquiry was conducted into methods of fermentation throughout cocoa-growing countries and it revealed a wide range of duration of fermentation from one and a half days up to ten days (Forsyth and Quesnel, 1957). The major difference lies in the variety of cocoa fermented, which has already been mentioned. The enquiry showed that Criollo cocoa is fermented for two to three days and Forastero cocoa for six to eight days, though some countries which formerly grew Criollo continue to use methods applicable to that type. While the duration of fermentation for different traditional methods is the same where the same type of cocoa is being fermented, the new tray method described later uses shorter fermentation periods.

Under-fermentation will produce beans with more purple pigment, and greater bitterness and astringency will be expected in the final product. Over-fermentation will produce beans with a dull dark-coloured nib and little chocolate flavour.

Turning   The purpose of turning the beans during fermentation is to ensure uniformity. Inevitably there are differences between one part of the fermenting mass and another so that turning is important to even out these differences. In box fermentation the wet beans will usually settle down into a solid mass during the first day, while the sweatings are draining off. Turning of this mass of beans is necessary in order to allow air to penetrate. There are many variations in the frequency of turning, from no turning at all to turning once or twice a day with fermentation in barrels which has been tried in the Ivory Coast and Cameroon. The commonest practices are turning every day or every other day, while some planters turn after one day and then every other day.

Seasonal effects   These influence the climatic conditions under which fermentation is undertaken and also the nature of the beans themselves. In Trinidad planters find that fermentation takes six days during the wet season and eight days during the dry (Anon, 1957); this is contrary to findings in West Africa where Allison and Kenten (1963) studied seasonal changes in some detail and found that temperature rose more slowly in June and July, the months of heaviest rainfall. There was some evidence that this was due to higher moisture content in the beans during June and July. The apparent contradiction between West Africa and Trinidad may be due to different moisture relationships.

These difficulties with fermentation may be countered. In Trinidad when the beans are unusually dry they are sprinkled with water or molasses. In West Africa it has been reported from Ikiliwindi that during the wettest months the wet beans are left for up to six hours in a large box before loading the trays (Anon, 1965).

The various methods of fermentation will now be dealt with in more detail.

Box fermentation

Fermenting boxes are made in a wide variety of shapes and sizes, but a typical one, measuring 1·2 by 1·2 m (4 by 4 ft) and 0·9 m deep (3 ft 6 in), will hold just over one ton of wet beans when loaded to a depth of 0·75 m (2 ft 6 in). Such boxes may be made individually or a row of boxes may be constructed as a single unit with partitions. They are made of hardwood and it is usual for the planks forming one or more sides to be slotted into the corner posts so that their removal will assist the movement of beans from box to box. The floor of the boxes is usually made of similar planks in which 15 mm holes are drilled at intervals of 10 to 15 cm (4 to 6 in). These holes are essential to provide drainage for sweatings and entry for air. In Trinidad the floor of the fermenting boxes is often constructed of slats spaced about 5 mm apart.

The size and number of boxes will depend on the size of harvests and the number of turnings given to the beans. The arrangement of the boxes is traditionally in a row; more recently boxes have been arranged in tiers either making use of a slope or raising initial boxes on a stand. The object of this has been to reduce labour involved in turning the beans but figures published by Allison and Rohan (1958) indicate that the saving afforded by this arrangement is only one man-hour per ton of dry beans. If these figures are typical, then the value of arranging fermenting boxes in tiers, which involves considerable extra capital cost, might be questioned.

The fermenting boxes, or sweat boxes, are usually housed in a separate building and given some protection from the weather. This will assist to provide insulation to the fermenting beans and hence maintain their temperature. As the beans produce a considerable volume of sweatings during the early part of fermentation, suitable drainage is often provided. The sweatings are acid and will attack cement or concrete, so that proper drainage tiles would be necessary for permanent use. This, however, is a council of perfection which is aesthetically attractive but unnecessary for the purpose of fermentation.

65. A set of fermenting boxes, with wooden shovel and rake, Venezuela.

66. A tier of boxes, Sabah.

67. (*left*) A fermentation tray.

69. 'Dancing' partially dried beans, Belize.

70. Apparatus for washing beans, Surinam.

71. A sun-drying platform with sliding roof, Trinidad.

72. A drying floor or 'tendal' in Ecuador. At night the beans are protected by the covers on the left.

In carrying out fermentation the wet beans are placed in the boxes to a depth no greater than 0·75 m. A greater depth reduces aeration and results in less even fermentation. It is important to ensure that the holes in the floor of the boxes or the gaps between the slats are clear prior to fermentation.

The wet beans are covered by banana leaves or sacks in order to retain the heat. Polythene sheet is also used but care should be exercised lest it prevent adequate aeration. The mass of beans must not be hermetically sealed.

During the course of fermentation the beans are moved from one box to another in order to ensure uniform conditions. It also helps to revitalise the fermentation as the temperature rises immediately after mixing. A typical box fermentation will last for six days, during which time the beans are mixed twice after two and four days; there are, however, variations on this pattern discussed elsewhere.

Heap fermentation

In heap fermentation the wet beans are placed on banana or plantain leaves on the ground. When the heap is complete, it is covered with more leaves and these are often held in place by small logs.

The method is used generally in West Africa but particularly in Ghana where farmers were recommended to ferment their beans in heaps for six days, turning after two and four days. Farmers' methods have not been surveyed but there is little doubt that the usual period is shorter and that turning may not be done at all. One reason for this is that in a well-fermented sample, the beans are larger, the shell being freer from the cotyledon. Thus the same weight of beans takes up more room than a less well-fermented sample which the Ghana farmer interprets as a greater loss in weight. This change in bulk density has not been measured, but it has been shown that the difference in loss in weight between fermented and unfermented beans is only 1·5 per cent so that the difference between partial and complete fermentation will be even less in this respect (West African Cacao Res. Inst., 1948).

Basket fermentation

Beans are frequently fermented in baskets in Nigeria. The baskets are of no definite size and are lined with leaves, which also cover the beans.

Tray fermentation

This relatively new method of fermentation evolved from Rohan's studies of heap fermentation (Rohan, 1958a). In examining heaps of different sizes he noticed that the surface layer of 3 to 4 inches was well fermented within two or three days, long before the remainder of the heap. From these observations the tray method in which 4-inch layers are fermented in trays stacked one on top of another was evolved (Allison and Rohan, 1958).

The trays are usually 0·9 m by 0·6 m by 13 cm deep (3 ft by 2 ft by 5 in). Battens are fixed across the bottom of the trays and some matting rests on the battens to hold the beans. Allowing for the battens and matting, the effective depth of the tray is 10 cm (4 in) and such a tray will hold 45 kg (100 lb) of wet beans.

The beans are loaded into the trays after pod opening and the trays are stacked up to twelve or fourteen trays high. Provided the depth of the beans is not altered there is no scientific reason why trays of different dimensions or stacks of greater or less height should be used, but trays of the above dimensions are convenient for two men to handle and a stack twelve high is about the maximum that labourers of ordinary height can manage.

The bottom tray is often left empty in order to improve aeration and the whole stack may be raised slightly for the same purpose and to allow sweatings to drain off. After twenty-four hours the stack of trays is covered with sacking in order to retain heat. The trays are not moved until the end of fermentation.

Fermentation in trays has been found to be quicker than fermentation in heaps or boxes, a four-day tray fermentation giving a product superior to normal heap fermentation. While this period may be adequate for Amelonado or Amazon cocoa in Ghana, it would require further testing with other varieties in other countries.

The tray method offers certain advantages over traditional methods. First, the fermentation period is shorter so that the throughput of beans is quicker. Second, as the beans are not turned during fermentation, there is a saving in labour which is of the order of 20 per cent. Third, with a stack of twelve trays there is an effective depth of 1·2 m (4 ft) of cocoa; thus less space is required. As initially described the trays were 1·2 by 0·9 m, half the tray being used for fermentation, after which the beans were spread out over the whole tray in a layer 5 cm (2 in) deep for sundrying. This adaptation of the method has not been found to be convenient for large-scale use but might be considered by smallholders. Several years' experience with the method in West Cameroon has shown its value on an estate as the individual trays form convenient units for moving beans, particularly when loading the dryers.

The appearance of the beans after tray fermentation is less uniform than after box fermentation. It is often found that the beans in the bottom tray and in the corners of the lower ones are covered with mould after fermentation but this has not been found to lead to any visible differences in the dried beans.

Judging the endpoint

By and large the fermentation of cocoa beans can be done by rule of thumb once a suitable method has been evolved to suit local circumstances. Nevertheless seasonal variations or different batch sizes may lead to changes in the progress of fermentation which would be revealed by temperature changes, colour, smell or internal appearance.

Temperature changes are not a good guide to the progress of fermentation as they will be influenced by other factors. Admittedly the temperature falls off as fermentation continues, but it rises after mixing and the temperature at the end of a normal fermentation period is probably the same as at some other point in fermentation. This is indicated by the temperature graph (Fig. 12.2).

At the beginning of fermentation the beans are pink-white and have a faint sweet smell. After the sweatings have run off the remaining pulp is dull white and gradually darkens to a red-brown colour. As the fermentation of the pulp proceeds, an acid smell develops and this is retained during the normal fermentation period. At the end of a six- or seven-day box fermentation the beans in the corners of the box have darkened further, becoming nearly black, and such beans will have an unpleasant ammoniacal smell. This marks the onset of changes associated with overfermentation and if a large proportion of the beans had this appearance and smell, then the batch as a whole would be overfermented.

The internal appearance of the beans is another guide to the progress of fermentation. Initially the beans (Forastero) will be bright violet with a white radicle or germ. After the death of the bean the space within the bean becomes filled with an exudate of a similar violet colour and this, together with the cotyledons and radicle, turns brown rapidly when the bean is cut open. At a later stage the exudate becomes a reddish-brown colour and the cotyledons become paler in the centre with a brownish ring around the outside. Such beans have been adequately fermented and are ready for drying. The difficulty in assessing the end point in fermentation is that the beans will not be uniform and it is only after trial and error that it is possible to decide the proportion of beans showing a brown ring that signifies adequate fermentation for the bulk. At River Estate, Trinidad, a proportion of 50 per cent of such beans was found to be appropriate (Quesnel, 1958).

Difficulties with fermentation

In most countries fermentation can be relied on to proceed normally according to the local pattern with perhaps some seasonal variation. In some countries, Papua–New Guinea in particular, fermentation does not always start properly, the temperature of the beans failing to rise or rising very slowly. These are referred to as 'dead ferments'. The reason for such failures has not been determined precisely but it has been suggested that it might be due to insufficient or unsuitable contamination by micro-organisms. Bridgland (1959) studied fermentation in New Guinea and evolved different techniques in order to overcome this problem. This new technique or 'interrupted fermentation' consisted largely of changes during the early stages of fermentation, the new method being as follows:

The pods are opened four days after harvest and wet beans are placed in a receiving box of no fixed dimensions but with adequate drainage. They remain in this box overnight and are then spread out on a resting floor at a depth of 2·5 to 4 cm for six to seven hours. During these two stages most of the sweatings will have drained off and much of the alcohol is lost during the resting phase. This reduces the level of acetic acid at later stages. After six to seven hours on the resting floor the beans are left overnight in a long shallow box and covered to retain the heat. On the following morning the beans are turned into a normal fermenting box and turned daily for the next four days.

This method was claimed to give more consistent results than any other method but comparisons over a period of two years of this method with standard seven-day box fermentation turning the beans every day, failed to show any advantage for the 'interrupted fermentation' (Anon, 1966). The latter requires more capital and the method is more laborious so it can only be recommended for those odd places where other methods fail.

A similar difficulty is described as a 'slimy' fermentation in which the temperature of the beans fails to rise normally and the pulp becomes slimy. Such fermentations have been reported from Jamaica and Trinidad while the 'dead ferments' of Papua–New Guinea may also be slimy (Quesnel, 1972). The two main causes for slimy fermentations are unripe pods and exposure of the wet beans to rain, both these factors will lead to low sugar content in the pulp and a less suitable base for the usual micro-organisms. Slimy fermentations should be avoided, but if they occur, sugar or molasses should be added to the fermenting mass; if fruit flies are absent from the fermentary, they can be attracted by hanging a stem of overripe bananas in it.

Cleanliness

Boxes or trays used for fermentation inevitably become encrusted with dried mucilage and this is liable to turn mouldy. While it is undesirable to allow this material to accumulate, on the other hand there is no need for great cleanliness. Fermentation depends on the activity of micro-organisms and one source of inoculum will be the box or tray in which the beans are to be fermented. It is, however, essential that holes or openings for drainage and aeration be kept clean.

Other practices associated with fermentation

Washing   The practice of washing beans between fermentation and drying is still employed in Sri Lanka, Java and Surinam. Washing removes any pulp adhering to the shell and thus reduces the shell percentage of the dried beans. The loss in weight is about 4 per cent and the shell percentage of beans from Sri Lanka and Java, for instance, is usually about 9 per cent as compared with 12 per cent for West Africa. The washed and dried beans have a bright clean appearance which is superficially attractive but the shells are brittle and do not afford a reliable protection against mould and infestation, and for this reason most manufacturers prefer unwashed cocoa.

While the cocoa markets expect washed cocoa from countries which traditionally prepare it, there is no case for other producers to consider the practice. The loss in weight and costs of washing are not likely to be offset by any increase in price. On the other hand, washing undoubtedly makes drying easier, but the extent of any economies that might result where drying is done artificially have not been measured.

Dancing and polishing   This practice is employed in Trinidad and some other West Indian countries in order to improve the appearance of the beans. Dancing is done after the beans are partly dried when they are hard but not brittle. The beans, which are spread out on the drying platform, are sprinkled with water to which some planters make some addition, for instance the juice from the crushed leaves of the Trumpet tree (Anon, 1957). The beans are then 'danced' or walked on by labourers for up to thirty minutes. This is done early in the day so that the beans receive a full day's drying subsequently.

Beans that are dried artificially can be polished in a circular bin in which ploughs rotate keeping the beans stirred. Rotary drum dryers can produce a similar effect. A polished appearance is expected of beans from some countries, but from the manufacturers' point of view it generally confers no benefit, unless shell percentage is reduced and this is uncertain. Some evidence that dancing improves the internal colour of the beans, reducing the percentage of partly-brown partly-purple beans, has been produced but this work was done in Brazil where the cocoa tends to be insufficiently fermented (Maravalhas, 1966).

Claying   This is of passing historical interest. In Venezuela and Trinidad beans were at one time coated with a dry red earth, which was claimed to prevent mould development and was supposed to be a hallmark of well fermented cocoa. When used properly the amount of clay added was only 2 per cent by weight, but the practice was abused by much heavier coatings and by its use to conceal poorly prepared cocoa. The practice was forbidden in Trinidad in 1923 (Shepherd, 1932).

Drying

At the end of fermentation the moisture content of cocoa beans is about 55 per cent and this must be reduced to 6 to 7 per cent for safe storage. The drying process is not simply a matter of losing moisture as the chemical changes taking place inside the bean during fermentation will continue during drying until halted by lack of moisture or inactivation of the enzymes by some other means.

The rate of drying varies greatly according to the method employed, but there are certain limits. If drying is too slow there is a danger that moulds will develop on the outside and may penetrate the shell; there is a further danger that off-flavours may develop. These dangers will be prevented if the beans become skin dry within twenty-four hours.

Can cocoa beans be dried too quickly? This is not an easy question to answer. The opinion is commonly held that fast drying will spoil the flavour and that it is essential to dry slowly for the first day or two. This point has been investigated with Amelonado cocoa in West Africa where cocoa beans were dried in as little as fourteen hours (Howat *et al.*, 1957b). The dried beans were made into chocolate which was compared with chocolate from sundried beans. No consistent difference in flavour was found. This finding must be qualified by the fact that the rapidly dried beans had a strongly acid smell and were found to contain more acid than sundried beans. This difference is obvious in the dried beans but full factory processing, including the process of conching, will reduce the acidity so that the chocolate has a normal flavour (Powell, 1958).

The conching process is not carried out in precisely the same way by all manufacturers so that its effect on acid beans will vary. Furthermore, raw beans are often judged by intermediaries who may criticise acid beans. Objections can, therefore, be raised against acid beans but the high degree of acidity that was found in the beans dried in fourteen hours is unlikely to occur in beans dried in the usual period of two to three days.

To sum up, artificially dried beans will be more acid than beans dried in the sun but artificial drying over the normal period of 24 to 72 hours is unlikely to give rise to beans with such a high degree of acidity that they will prove difficult to market. Drying is done in the sun where quantities are small and the climate allows or by some form of artificial dryer for large production or adverse weather conditions.

Sundrying

In countries where the main harvest is gathered during the dry season the beans are usually dried in the sun. This applies to West Africa and the West Indies. In West Africa the beans are dried on mats raised off the ground or on concrete floors. The latter are less satisfactory as they have to be fenced to exclude domestic animals and poultry and because they tend to be dusty. In the West Indies, drying takes place on wooden drying floors with moveable roofs or, alternatively, on wooden trays on rails which can be pushed under a fixed roof.

In Trinidad, the area required for sundrying is at the rate of at least 2·8 m² per 50 kg (30 ft² per 110 lb) of dry beans, the normal drying floor being 18 by 6 m and having a load of 2,250 kg (44 cwt) dry beans. On such floors, the beans are spread out to a depth of about 5 cm (2 in) and are raked with a wooden palette so that the beans are in ridges with the drying floor exposed between them. By frequent movement of the ridges, the drying floor is kept relatively dry and the beans more uniformly exposed to the sun. Sundrying requires constant attention, not only to ensure uniform drying, but also to put the beans under cover when it rains.

In West Africa, the drying mats are erected in the villages where the beans can be spread out in the morning, turned and cleaned during the day and heaped at night, or in the event of rain.

The length of time it takes to dry in the sun depends obviously on the weather. It is unusual for drying to be completed in less than a week, but during dull weather, the period may extend to two weeks or more.

During the course of sundrying, it is customary to pick over the beans, removing pieces of pod husk, placenta, foreign matter and defective beans and also to separate beans which stick closely together.

Artificial drying

Conditions for drying   A number of factors are involved in finding the most economical conditions for drying; they include depth of beans, rate of airflow, hot air temperature. The effect of these factors was studied by Shelton (1967) who conducted a series of small-scale experiments with a specially designed apparatus. Within the limits of these experiments it was shown that a deep layer of 25 cm, a low air-flow of 3 m per minute and a moderate temperature of 60 to 65°C gave the most economical conditions. While these figures are a guide to the most economical conditions, they must be modified by the type of dryer being used. This applies particularly to rate of airflow. While a low rate may dry the beans with greater efficiency, the heat content or entropy of the hot air is small and a significant proportion may be lost in the ducting and plenum chamber before passing through the beans. Less may be lost by a lower temperature and higher rate of airflow so that more useful heat reaches the beans. The most economical conditions for a dryer would be subject to experiment, but the above results serve as a guide.

There is another school of thought regarding the most economical conditions for drying. Salz (1972) favours 'high airflows (50 to 60 ft/min) at low temperatures, for the pre-drying stage, and low airflows

(30 to 35 ft/min) at high temperatures (180 to 185°F [82 to 85°C]) for the final stages of drying'. This view is based on experience in Papua–New Guinea, but the conditions quoted do not appear to have been selected as a result of experimental work.

Interrupted drying   In Papua–New Guinea a practice of interrupting drying has come into favour. The beans are dried for twelve hours, rested for twelve hours and then finished. The purpose of this resting period is to allow the moisture within the beans to migrate to the surface; it is also said to reduce acidity and prevent case-hardening of the bean. There seems to be a possibility of reducing fuel consumption by this means but there is no evidence on this point or any of the other assumed benefits. On the other hand this interruption results in the beans occupying the dryer for a longer period than with continuous drying; alternatively, the beans have to be moved to a separate bin for the resting period. This means more equipment and more handling.

The effect of interrupted drying on quality is untested. Keeping the beans for some time when still moist and cool will ultimately lead to mould development and the timing of the resting period entails a hazard to be borne in mind.

Heat exchangers or direct-fired heaters   The platform dryers described later employ heat exchangers or direct-fired heaters. Heat exchangers keep the hot air separate from the products of combustion but are more expensive in initial cost than direct heaters in which the products of combustion mix with the hot air, and the latter use 20 per cent less fuel so are cheaper to run. The use of heat exchangers has been favoured because it avoids the possibility of taint or toxic residue, but trials have failed to show that direct heaters cause any such defect, even when badly adjusted.

Artificial dryers

Simple dryers   This category includes dryers using wood as fuel and dependent on conduction or convection to heat the beans. Most other types burn oil and have some degree of control over airflow and temperature.

In Cameroon, large slate dryers were built many years ago; they are also used on estates in Fernando Po. These dryers have a drying platform built of slates and this is heated from a wood fire at one end of the chamber beneath the platform. Smaller dryers working on the same principle, but with one or two flues made of oil drums covered with sand and a drying mat, or with the flues embedded under a cement drying platform, have been built in West Cameroon and are known as 'Cameroons dryers'. These dryers are unsatisfactory because heat

73. A slate dryer with mechanism for stirring the beans, Fernando Po.

74. A Samoan dryer, Cameroon. Note the gap around the flue to allow air to enter.

75. Samoan dryer with sliding roof and fan in side wall, Sabah.

76. An ASP Universal dryer with sliding roof, designed for sun or artificial drying, New Guinea.

77. A drying platform connected to a heat exchanger unit. Tiers of boxes and trays in the background.

78. A rotary dryer, Venezuela.

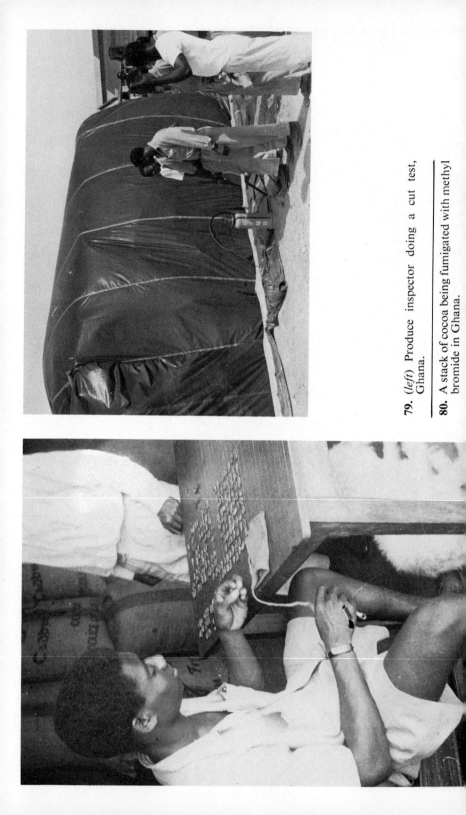

79. (*left*) Produce inspector doing a cut test, Ghana.

80. A stack of cocoa being fumigated with methyl bromide in Ghana.

distribution is not uniform and because it is easy for cracks to occur in the drying platform which allow smoke to escape.

The 'Cameroons dryer' has been replaced by the Samoan dryer which dries the beans by means of a convection current. This dryer consists of a simple flue in a plenum chamber and a permeable drying platform above. Air inlets must be provided in order to allow the convection current to flow and the flue must be properly sealed to prevent the escape of smoke (Cadbury Brothers, 1963).

This is basically a very simple dryer which can readily be constructed out of materials that are easily available. The materials can be altered to suit local conditions and the dimensions can be changed within certain limits. The width of the drying platform is governed by the ease of mixing; the length of the drying platform will affect the uniformity of drying as there is usually more heat at the end next to the fire. If the platform is longer than 6 m, the difference in temperature may result in an excessive difference in drying speed between one end and the other; it is best to restrict the length to 5·5 to 6 m (18 to 20 ft). Some planters have installed electric fans in the walls of the plenum chamber; these fans blow air into the plenum chamber, thereby increasing the rate of drying, and ensuring greater uniformity over the drying platform.

A Samoan dryer with a drying platform 3·5 × 3 m will have a capacity of 225 to 275 kg (500 to 600 lb) dry beans per batch and the drying period should be one and a half to two days, but this will vary considerably according to the fire and the temperature. The amount of fuel consumed has been estimated to be one and a half tons of firewood per ton of dry beans. Again this figure is liable to vary considerably.

The Samoan dryer was a small, simple version of the Martin dryer, which is used on certain estates in Western Samoa. The Martin dryer has a drying platform 13·5 m long and 7·5 m wide (15 by 8 yd) and is fitted with a U-shaped flue. The roof runs on rails so that sundrying can be done. The capacity is 3 tons of dry beans and the drying period is twenty hours.

In Bahia, cocoa estates use dryers of two types. The 'secador' has a single platform and the 'estufa' a number of trays. In both cases a convection current of hot air is generated by a flue.

All these non-mechanical dryers are cheap to build and may be cheap to run if wood fuel is easily available. The Samoan dryer is likely to be suitable for small farmers or for small co-operative societies, and have been built for this purpose in the Solomon Islands. The drawback to these types is that the flues must be carefully maintained. The cheap and simple flue will corrode fairly quickly and this can lead to smoke escaping and contaminating the beans.

On large estates and plantations, drying in the sun or by use of types of dryer described above may be more costly than other means of artificial drying. The drying period is longer, particularly for sundrying, and therefore, the area required for drying is much more extensive. Further, the labour required to handle large quantities when sundrying becomes considerable.

Platform dryers   The first dryer of this type to be used extensively was the ASP 'Universal' dryer used in New Guinea and designed to combine sundrying with artificial drying. The usual model had a drying platform 12 by 6 m with a wire mesh floor raised about 0·5 m off the ground. The platform was covered by a sliding roof running on rails. For artificial drying, the unit was provided with an oil-fired heat exchanger and fan driven by a diesel engine. The capacity of the unit was 2·5 tons dry cocoa per batch which could be dried in less than two days.

It is difficult to combine sundrying with artificial drying in one unit, as the optimum depth of beans for the two methods is markedly different. When sundrying, the beans are spread in a relatively thin layer, but on a drying platform through which hot air is blown, the beans should be placed in a deep layer in order to make the most economical use of the heat available.

Platform dryers for artificial drying are simple in principle but usually have a number of controls in order to maintain the desired temperature and provide safety devices. Typical installations of this type have been described by Wood (1961) and by Allison and Kenten (1964). The drying platforms are long and narrow to ensure uniform distribution of the hot air and ease of handling and mixing the beans, a typical installation having dimensions 9 by 2·5 m (30 ft by 8 ft 6 in). The drying platform can be made of nylon-covered expanded metal or of perforated aluminium sheet. If perforated sheet is used the holes should be close together, about 3 cm between centres and the holes or spaces should not exceed 1 cm in diameter or length. The plenum chamber beneath the platform is about 1 m deep and is connected to the fan and heater by a fishtail duct. The fan and heat exchanger unit provide a certain volume of air at a predetermined temperature; the volume may be controlled by means of a damper. The whole unit requires a supply of electricity.

Dryers of this type make use of machinery designed for many purposes so there is a variety of sizes and capacities available. In the trials first reported, a drying platform 6 by 2·7 m was coupled to a fan and heat exchanger unit capable of providing 70 m³ per min (2,500 cfm) at a temperature 50°C above ambient. The trials showed that a high temperature and low airflow coupled with a deep layer gave the most economical conditions, the fuel consumption being about 180 to 230 l (40 to 50 gal) per ton of dry beans. The capacity of the dryer was 1 ton of dry beans and drying period thirty-two to thirty-four hours.

Similar results were obtained at Tafo by Allison and Kenten (1964). In this case the dryer was installed in an enclosed space and it was possible, by means of a flap, to recirculate the air within this space or draw in outside air. At the beginning of the drying period the flap was open, air was drawn in and could get out through windows and doors; as drying proceeded the windows, doors and flap were closed to re-circulate the air. With this arrangement it was possible to produce 1 ton dry beans within twenty-four hours at a fuel consumption of 155 l (34 gal) per ton.

It is often found that the output and fuel consumption achieved in trials cannot be repeated in practice on estates and at Ikiliwindi in West Cameroon, experience over several years led to some changes in method. The depth of beans was increased to nearly 30 cm (12 in) in the initial stage and during the busy season when several dryers are loaded at the same time, the beans from two dryers are transferred to a third after twenty-four hours. By this means, batches up to 6·5 tons have been dried in sixty hours at a fuel consumption of 160 l (35 gal) per ton. With smaller batches fuel consumption is higher.

Similar findings have been reported from Sabah where over 2 tons dry beans have been produced in thirty hours at a fuel consumption of 114 l (30 gal) per ton (Wyrley-Birch, 1968). To achieve this output the beans were loaded to a depth of 25 cm along the sides of the platform and heaped up to 36 cm in the middle.

To obtain good fuel economy it is essential to load the platforms as deep as possible. This raises the problem of adequate mixing of the beans in order to get uniform drying and to prevent the beans sticking together. Mixing can be done manually up to a depth of 25 cm but beyond that it becomes increasingly difficult to do the job effectively. Mechanical agitators have been designed but are not in widespread use. There are problems in designing an agitator which will mix the beans properly without leaving pockets of unstirred beans and without excessive breakage. These problems would be reduced with a circular drying platform but it is possible that other problems would arise.

Bin dryers   In Papua–New Guinea the drying platform has been replaced by a drying bin of similar dimensions but holding beans up to a depth of 45 cm. Some of these bins are sealed or covered and arranged so that the airflow can be reversed. This helps to overcome the lack of uniformity that will occur in a deep layer but the cocoa must still be stirred and this is not easily done (Salz, 1972).

Rotary dryers   It is generally conceded that rotary drum dryers are not suitable for the complete drying of cocoa beans, though they can be used for the last stage of drying. There are two main reasons for this: first, the sticky remains of the mucilage on the beans tends to clog the holes in the drum; second, the mass of beans shrinks during the course of drying to about two-thirds its original volume. In order that the wet beans should move and mix, it is essential that some free space be left when the dryer is filled. As the beans dry and shrink the drum will become about half full and this will lead to considerable loss of hot air which will not pass through the beans.

Rotary dryers are used for the later stages of drying after the beans are externally dry. Rotary dryers are used on various crops but for cocoa there are certain important points of detail which have been discussed by Newton (1966). First, the drum should revolve slowly at about 0·25 to 0·5 rpm in order to prevent breakage of shell and bean. Second, there should not be any internal baffles as they will also cause breakage. Finally, the holes in the metal skin of the drum should be 8 to 10 mm in diameter.

The use of a rotary dryer will polish the beans which is considered an advantage in some sections of the cocoa market. Polished beans give the appearance of having the adhering mucilage removed but an investigation gave a higher shell percentage on beans that had been dried and polished in a rotary dryer as compared with sundried beans (Knapp, 1937). This indicates that the mucilage is spread uniformly rather than removed from the beans.

Other dryers. Lister Moisture Extraction Unit. This unit, consisting of a diesel engine and fan, produces a large volume of air at 6 to 12°C above ambient. It is used for drying hay, grain and other crops. Trials were conducted with cocoa beans in New Guinea (Newton, 1963, 1965). The large volume of air up to 990 m³ per min (35,000 cfm) allows large quantities of beans, up to 5 tons dry weight, to be dried in one batch but the low temperature results in a lengthy drying period and in the trials foul odours developed. In order to overcome this, it is necessary to get the beans to an externally dry condition within twenty-four hours and this can be done by means of a shallow tray for initial drying and a deep bin for the later stages. The manufacturers have designed a combination bin for this purpose and have added a direct-fired heater to the unit to raise the temperature of the hot air. The advantage of this type of dryer, which has not been fully tested, is its lower capital cost. Published results suggest that fuel consumption is similar to other dryers but no additional supply of electricity is required.

Tunnel dryers are in use in Surinam where the beans are washed. The beans are spread in a thin layer on trays and the trays are loaded on to a trolley which is placed in the tunnel. Hot air is blown through the tunnel and arrangements are usually made for recirculation.

Buttner dryer. This is the only continuous dryer that has been used for cocoa beans. It consists of an endless belt of small trays, each holding about 40 kg wet beans. The trays pass down a drying tower through which hot air is blown. The beans are dried in sixteen hours and the machine can handle 9 tons wet beans when fully charged.

Moisture content and testing

For safe storage the moisture content of cocoa beans should be between 6 and 7 per cent; above 8 per cent there is danger of moulds developing within the beans; below 5 per cent the beans will be very brittle. It might seem that the judging of moisture content within such a narrow range would be difficult but in fact the task is relatively simple when testing beans that have been in store for some time. If the beans are in an open bag the feel and rattle of properly dried beans when the hand is thrust into the bag is characteristic and can be learnt with a little experience. If a bean is squeezed between the fingers it should not bend or shatter, but should break into two pieces.

Judging the end point of the drying process is more difficult, particularly when the beans are at a high temperature as this may cause them to soften. When drying artificially, small samples should be taken and cut when they have cooled. The half beans should then be broken, and should break into two parts rather than bend or shatter. Such samples are unlikely to be uniform, some being properly dry, others still rather damp. It is probably unnecessary to continue drying until all the beans in a sample pass this test, as drying will continue for some time as the beans cool down. Experience will tell the stage at which to stop the drying process, which unfortunately cannot be run to a strict routine, as drying times vary.

Moisture meters cannot be used for this particular purpose because of the uneven distribution of moisture both within and between beans at the end of drying. They can however, be used for testing beans in store although the hand testing and cut test is usually adequate for this purpose. There are several moisture meters available that give an instant reading of moisture content and are based on a measurement of electrical conductivity. While convenient to use, the instruments have to be calibrated before use and their accuracy will require checking at intervals. The readings may be affected by changes in ambient temperature and by pressure between the beans—beans in a stack may give a different reading from beans in a free standing bag due to pressure within the stack. The meters that have been used are the Scot Mec-Oxley which has nine electrodes and a built-in generator actuated by a handle; the KPM meter with two electrodes, and the Marconi moisture meter for which the beans have to be ground and placed in a small cell in which they are compressed.

Cleaning and bagging

After drying, the beans will be bagged in jute sacks and, if still hot, allowed to cool. At this stage it may be necessary to improve the quality of the sample by removing flat and broken beans. The incidence of flat beans varies according to planting material and growing conditions; an examination of a sample will tell whether any cleaning or grading is necessary. Where grading of this nature is needed, it can be done with a grading machine with reciprocating sieves or a rotating drum; the former type of machine often incorporates a fan which blows away any dust and small pieces of shell.

The cleaned sample will consist of whole beans with a fairly wide range of size. Further grading according to size is not considered to be worth while, as there is no demand for beans of different sizes, nor at present is there any strong emphasis on uniformity of bean size.

The next stage is to complete the bagging and weighing. It is recommended and frequently stipulated in grading regulations that new jute sacks be used when cocoa is exported. In many countries a cocoa sack will hold 140 lb (63·5 kg) beans net weight, but in countries using metric weights, a close equivalent, e.g., 67 kg, is used. Trinidad and Grenada use larger bags holding 165 and 200 lb (75 and 90 kg) beans respectively.

Storage

Cocoa beans can be stored indefinitely in temperate countries but in the tropics the high temperatures favour the rapid development and spread of stored product pests and the humidity may be high enough to allow development of moulds. If cocoa is to be stored for any length of time in the tropics special precautions must be taken to ensure that quality does not deteriorate from either of these causes.

A cocoa store should be built with these requirements in mind. To prevent infestation the building should have a cement floor and walls of brick or concrete blocks; wooden floors and walls provide cracks and crevices in which stored product pests can hide and breed.

Cocoa beans are hygroscopic and will absorb moisture under very humid conditions. The following table shows the equilibrium moisture content at various relative humidities.

Relative humidity	Moisture content
75	7·3
80	7·7
85	8·7
90	11·6
95	15·5

Cocoa beans with a moisture content in excess of 8 per cent will turn mouldy. Therefore the relative humidity in cocoa stores should not exceed 80 per cent for any length of time. The relative humidity in the open air exceeds this figure in many tropical countries for a large part of the day. Inside a cocoa store the relative humidity will be somewhat lower and it has been suggested that proper construction and operation of a store will help to maintain lower relative humidities (Powell and Wood, 1959a). Such improvements can be affected by the use of insulating material on the underneath of the roof and by keeping the store closed as far as possible during the early hours of the day. These measures are only necessary where relative humidities are exceptionally high as in Cameroon. In other parts of West Africa, humidities are slightly lower and studies of storage conditions have shown that special measures are not necessary.

Experiments in West Cameroon showed that dry beans at 4·5 per cent moisture content in a free-standing bag can absorb moisture up to the danger level of 8 per cent in three weeks (Powell and Wood, 1959b). This is the extreme case; stacks of cocoa would not, on average, absorb moisture at this rate although bags on the edge of a stack may approach this condition. In Nigeria it has been found that the temperature in large stacks rises several degrees above ambient. This leads to a fall in moisture content in the inner bags, but the outer bags can gain moisture slightly. Thus large stacks can maintain a suitable environment provided air can circulate freely around them. Any impedance to free circulation by covering the stack with a tarpaulin, for instance, could lead to damp conditions and mould development at the top of the stack.

In most parts of the tropics cocoa beans can be stored safely from the point of view of mould development for the usual interval of time between bagging and shipment. Where particularly humid conditions occur, as in West Cameroon, or where the interval between bagging and shipment is likely to be prolonged, the beans can be protected against the intake of moisture and consequent danger of mould development by the use of polythene liners inside the normal jute bag. Polythene is impervious to moisture and will prevent the uptake of moisture by dry beans; on the other hand, beans which are not properly dry cannot lose moisture if packed inside a polythene liner and such beans are likely to become mouldy. It is, therefore, essential that beans packed in liners should be dried thoroughly to 6 or 7 per cent moisture content. The liners can be quite thin (150 gauge is adequate), but they must be larger than the bag in which they are placed. Liners 1·5 by 0·75 m (5 ft by 2 ft 6 in) are of suitable size for the normal jute bag which holds 140 lb (63·5 kg) beans. When the bag and liner have been filled, the neck of the liner can be tied or simply folded two or three times.

References

Allison, H. W. S. and Kenten, R. H. (1963) 'Seasonal variation in the fermentation of West African Amelonado cocoa', *Trop. Agric.*, **40**, 217–22.

Allison, H. W. S. and Kenten, R. H. (1964) 'Mechanical drying of cocoa', *Trop. Agric.*, **41**, 115–20.

Allison, H. W. S. and Rohan, T. A. (1958) 'A new approach to the fermentation of West African Amelonado cocoa', *Trop. Agric.*, **35**, 279–88.

Alvim, P. de T. (1966) 'Factors affecting flowering of the cocoa tree', *Cocoa Growers' Bull.*, **7**, 15–19.

Alvim, P. de T., Machado, A. D. and Vello, F. (In press) 'Physiological responses of cacao to environmental factors', *Proc. 4th Int. Cocoa Res. Conf., Trinidad, 1972*.

Anon (1957) *Preparation of cocoa*. Publ. Exch. Serv. No. 60, Caribbean Comm. Trinidad.

Anon (1965) 'Tray fermentation at Ikiliwindi', *Cocoa Growers' Bull.*, **5**, 25.

Anon (1966) 'Processing and marketing of cocoa in Papua and New Guinea', *Tech. Mtg Cocoa Prod. Honiara 1966*, S. Pacific Comm. Paper 39.

Bridgland, L. A. (1953) 'Study of the relationship between cacao yield and rainfall', *Papua and New Guinea Agric. Gazette*, **8** (2), 7–14.

Bridgland, L. A. (1959) 'Processing methods for cacao growers in Papua and New Guinea', *Papua and New Guinea Agric. J.*, **12**, 87–115.

Cadbury Brothers (1963) *The Samoan Cocoa Drier*, 3rd edn, Bournville.

Cooper, St. G. C. (1940) 'A note on the maturation period of cacao pods in Grenada', *Trop. Agric.*, **17**, 165.

Forsyth, W. G. C. and Quesnel, V. C. (1957) 'Variations in cacao preparation', *Proc. 6th Inter-American Cocoa Conf., Salvador 1956*, pp. 157–68.

Haworth, F. (1953) 'Continuous cacao fermentation temperature records at San Juan Estate 1949', *Rep. on Cacao Res. 1945–51*, Trinidad, p. 109.

Howat, G. R., Powell, B. D. and Wood, G. A. R. (1957a) 'Experiments on cocoa fermentation in West Africa', *J. Sci. Food Agric.*, **8**, 65–72.

Howat, G. R., Powell, B. D. and Wood, G. A. R. (1957b) 'Experiments on cocoa drying and fermentation in West Africa', *Trop. Agric.*, **34**, 249–59.

Jabogun, J. A. (1965) 'A mechanical cocoa pod-opener', *Nig. Agric. J.*, **2** (1), 44.

Jimenez, S. E. (1967) 'Zinke Pod-Breaker—A significant contribution to the development of the cacao industry', *Cacao*, **12** (2), 1–5.

Kenten, R. H. and Powell, B. D. (1960) 'Production of heat during fermentation of cacao beans', *J. Sci. Food Agric.*, **11** (7), 396–400.

Knapp, A. W. (1926) 'Experiments in the fermentation of cacao', *J. Soc. Chem. Ind.*, **45**, 140–2.

Knapp, A. W. (1937) *Cacao Fermentation*, London, Bale Sons & Curnow, p. 149.

MacLean, J. A. R. (1953) 'Some chemical aspects of black pod disease in West African Amelonado cacao', *Emp. J. exp. Ag.*, **21**, 340–9.

MacLean, J. A. R. and Wickens, R. (1951) 'Small-scale fermentation of cocoa', *Rep. Cocoa Conf. London 1951*, pp. 116–22.

Maravalhas, N. (1966) 'The effect of "dancing" on the quality of fermented cocoa', *Trop. Agric.*, **43**, 351–4.

McKelvie, A. D. (1958) 'Frequency of harvesting', *Ann. Rep. W. Afr. Cocoa Res. Inst. 1956–57*, p. 63.

Newton, K. (1963) 'Cocoa drying with the Lister Moisture Extraction Unit', *Papua and New Guinea Agric. J.*, **16**, 91–102.

Newton, K. (1965) 'Cocoa drying with Lister Moisture Extraction Unit. Addendum: Drying trial on an 800 sq. ft floor', *Papua and New Guinea Agric. J.*, **17**, 109–16.

Newton, K. (1966) 'Notes on cocoa drying in Papua and New Guinea', *Tech. Mtg on Cocoa Prodn., Honiara 1966*, S. Pacific Comm., Paper 18.

Powell, B. D. (1958) 'The rapid artificial drying of cocoa beans and chocolate flavour', *Trop. Agric.*, 35, 200–4.

Powell, B. D. (1959) 'Cocoa fermentation: A note on the anaerobic and aerobic phases', *Chem. & Ind.*, pp. 991–2.

Powell, B. D. and Wood, G. A. R. (1959a) 'Storage, transport and shipment of cocoa, 1. Prevention of moulding during storage', *World Crops*, 11, 314–16.

Powell, B. D. and Wood, G. A. R. (1959b) 'Storage, transport and shipment of cocoa, 2. Protection during handling and shipment', *World Crops*, 11, 367–8.

Quesnel, V. C. (1958) 'An index of completion of the fermentation stage in cacao curing', *Proc. 7th Conf. Inter-Amer. de Cacao, Palmira 1958*, pp. 512–16.

Quesnel, V. C. (1972) 'Slimy fermentations', *Cocoa Growers' Bull.*, 18, 19–23.

Rohan, T. A. (1958a) 'Observations on the fermentation of West African Amelonado cocoa', *Proc. Cocoa Conf. London 1957*, pp. 203–7.

Rohan, T. A. (1958b) 'Processing of raw cocoa, II', *J. Sci. Food Agric.*, 9, 542–51.

Rohan, T. A. (1958c) 'Cocoa preparation and quality: fermentation', *Rep. W. Afr. Cocoa Res. Inst. 1957–58*, pp. 60–3.

Salz, A. G. (1972) 'Cocoa processing—A practical approach to fermenting and drying', *Cocoa and Coconuts in Malaysia, Proc. Conf. Incorp. Soc. Planters, Kuala Lumpur 1971*, pp. 181–218.

Shelton, B. (1967) 'Artificial drying of cocoa beans', *Trop. Agric.*, 44, 125–32.

Shepherd, C. Y. (1932) *The Cacao Industry of Trinidad. Some economic aspects. Part V, Historical 1921 to 1932*, Port-of-Spain, Trinidad Govt. Printing Office.

Toxopeus, H. (1964) 'F$_3$ Amazon cocoa in Nigeria', *Ann. Rep. W. Afr. Cocoa Res. Inst. (Nigeria) 1963–64*, pp. 13–23.

Wadsworth, R. V. (1953) Quoted in discussion, *Rep. Cocoa Conf. London 1953*, p. 42.

Waters, H. B. and Hunter, T. (1929) 'Measurement of rate of development of cacao pods', *Yearbook 1928 Gold Coast Dept. Agric.* Bull. 16, 121–7.

West African Cacao Research Institute (1948) 'Cacao fermenting experiments', *Ann. Rep. W. Afr. Res. Inst. 1946–47*, p. 67.

Wickens, R. (1955) 'Frequency of picking experiments', *Ann. Rep. W. Afr. Cocoa Res. Inst. 1954–55*, p. 55.

Wood, G. A. R. (1961) 'Experiments on cocoa drying in the Cameroons', *Trop. Agric.*, 38, 1–11.

Wood, G. A. R. (1968) 'Pod breaking machines. Progress report', *Cocoa Growers' Bull.*, 11, 15–26.

Wyrley-Birch, E. A. (1968) 'Processing cocoa at the Cocoa Research Station, Sabah', *Cocoa Growers' Bull.*, 11, 18–24.

A fuller treatment of this subject is to be found in:

Rohan, T. A. (1963) *Preparation of Raw Cocoa for the Market*, FAO Agric. Studies No. 60, Rome.

Chapter 13

Quality and Inspection

The quality of a sample of cocoa beans is judged primarily on the flavour of the chocolate made from them but is also dependent on other factors, such as bean size, shell percentage, fat content and number of defective beans. Cocoa of good quality will have the inherent flavour of the type of bean concerned together with the relevant physical characteristics and freedom from defects. The physical characteristics and the presence of defects can be measured objectively, but flavour is a matter for subjective judgment and can only be assessed objectively in so far as the presence of certain defects influences flavour.

Chocolate flavour is developed in two stages: the first involves the two processes of fermentation and drying on the farm, the second is the roasting process in the factory. On the farm, the final chocolate flavour is influenced by the type of tree from which the beans are harvested and the fermentation and drying processes.

In the case of many of the principal bulk cocoas the intrinsic flavour of the beans is comparatively uniform, and these cocoas are produced from Amelonado and Amazon cocoas. There are other types of tree —Trinitario and Criollo—which have different intrinsic flavours and these can produce 'fine' cocoas, fetching prices which may be similar to or higher than those for bulk cocoas. These differences of flavour from different types of tree can only be judged by the flavour of the finished chocolate (Wood, 1973).

Fine grade cocoas may be produced from Criollo or Trinitario beans and depend for their premiums on having particular flavour characteristics and having them uniformly from one consignment to another. Criollo cocoas have cotyledons which are white when fresh and pale cinnamon colour after fermentation and drying, and will produce chocolate with a mild nutty flavour. Such cocoas may be produced from pure Criollo trees or from some Trinitarios.

The word Trinitario is less frequently used to describe commercial types, the fine grades produced in Trinidad and the various other countries growing Trinitario cocoa being described by country of origin. Such cocoas have a full chocolate flavour and some fruitiness which is not normally found in West African cocoa. The cocoas from West Africa and Brazil, often called 'bulk' cocoas, which merely indicates their preponderance on the market, differ according to some of their quality characteristics, but all produce a good chocolate flavour.

In the past, bulk cocoas were generally of poor quality, while 'fine' cocoas were generally of good quality, so the two terms, quality and flavour, have sometimes been used as if they were synonymous. They are not, and it is important to be clear as to the distinction between them.

The relationship between flavour, quality and normal methods of examination has been described in *Raw Cocoa: manufacturers' quality requirements* (Cocoa, Chocolate and Confectionery Alliance, 1968):

> The chocolate manufacturers' main criterion of quality of all types of beans is the flavour which is developed by roasting and processing them. Flavour, however, is essentially a subjective judgment by the taster, and cannot be made into an objective measurement to enable the limits of satisfactory flavour to be specified on paper and used for grading cocoa in normal commercial operations.
>
> The flavour of some bulk cocoas is comparatively uniform when the beans are properly prepared for the market by the traditional methods. A useful assessment of the quality of such cocoas can be made by a physical examination of the beans, and especially their appearance when cut. The criteria used in the cut test are more readily expressed objectively than flavour, and consequently have attracted more emphasis in specifying the manufacturers' quality requirements in these bulk cocoas than their intrinsic importance warrants.

Physical characteristics

There are aspects of quality of importance to manufacturers over which the producer has little influence; these are bean size, shell percentage and fat content.

Bean size is of importance to the manufacturer because it affects shell percentage, fat content and the initial roasting process. Manufacturers want to buy beans with the lowest shell percentage consistent with adequate protection of the nib from mould and insects and with the highest fat content. These two factors do not vary with bean size, but, in beans weighing less than 1 gm (0·04 oz), shell percentage increases and fat content is reduced. The following figures from samples of

Ghana beans illustrate the point:

Bean size	Shell (%)	Fat in nib (%)
Below 1 gm	14·03	55·86
Above 1 gm	11·41	57·63

There is little change in shell or fat percentage as bean weight increases above 1 gm so, from the manufacturers' point of view, there is no merit in trying to produce especially large beans. Therefore, manufacturers have stated that the average weight of dried beans should not be less than 1 gm. Bean size is frequently measured in West Africa on the basis of number of beans to 11 oz and the equivalent to 1 gm per bean is 312 beans to 11 oz. Nigeria is the only country selling two grades according to bean size, purchases with more than 300 beans to 11 oz being graded as 'Light crop', as distinct from 'Main crop' for other purchases.

The effect of bean size on the roasting process is a more direct one, small beans being roasted more quickly than large ones. For this reason a certain uniformity of bean size is desirable, and it has been suggested in the International Cocoa Standards that not more than 12 per cent of the beans should have bean weights outside plus or minus one-third of the average bean weight. Currently, range of bean size is rarely determined, but should new planting material lead to greater spread of bean weights, this factor could become of greater importance.

The shell percentage of beans from Ghana and Nigeria is 11 to 12 per cent and this is generally accepted as the standard. Many cocoas, particularly Trinitarios, have shell percentages well above 12 per cent; beans from Trinidad and New Guinea, for instance, have 15 to 16 per cent shell. As shell is of very little economic value once it is separated from the nib, manufacturers have a preference for beans with shell percentage at the West African level.

Fat content is determined by extraction with a solvent and is commonly expressed as a percentage of the dry nib or cotyledon. In West African beans the fat content is usually 56 to 58 per cent. With Forastero cocoa the figure can vary from 55 to 60 per cent; Criollo beans usually have a lower fat content of about 53 per cent.

Statistics of bean size, shell and fat percentage of beans from various countries gathered from the 1961/62 main crop are given in Table 13.1. In this table, shell and germ are combined, which raises the figure by nearly 1 per cent, as the germ forms 0·8 to 0·9 per cent of the whole bean.

The table indicates the higher shell content from certain cocoas, especially Papua–New Guinea and Trinidad, and the low fat percentage in Criollo beans from Mexico and Indonesia. The low shell percentage in the latter samples indicates that they have been washed.

Further details of Ghana and Nigerian beans are given in Table 13.2 which shows the variations from year to year.

Table 13.1 *Analysis of cocoa beans from various countries*

Country	Weight of 100 beans	Shell and germs %	Fat %
Africa			
Ghana	105·8	13·3	56·9
Nigeria	113·8	12·3	57·3
Ivory Coast	112·8	12·6	58·2
Cameroon	101·8	12·9	56·3
Latin America			
Brazil	104·0	15·3	56·5
Ecuador	129·0	14·2	53·4
Trinidad	105·7	17·3	56·7
Mexico	113·4	9·0	53·2
Grenada	97·9	16·6	56·7
Asia			
New Guinea	120·4	16·4	56·9
Indonesia	105·3	9·9	53·6

SOURCE: *Cocoa Bean Tests 1961–62*, Gordian, Hamburg.

Table 13.2 *Analysis of Ghana and Nigeria maincrop beans*

	Ghana				Nigeria			
	1966–67	1967–68	1968–69	1969–70	1966–67	1967–68	1968–69	1969–70
Weight of 200 beans in grams	213	211	226	210	219	218	219	221
Butter fat in dry nib (per cent)	57·5	57·5	57·8	57·7	57·1	56·7	56·5	57·2
Shell (per cent)	11·8	11·9	11·6	12·4	12·2	12·2	12·5	12·9

Factors affecting physical characters

Bean size, shell and fat percentage are all influenced by the distribution of rainfall during the period of development of the crop. A close correlation between rainfall and bean weight, and probably other pod and bean values, has been established by a detailed study carried out in Nigeria (Toxopeus and Wessel, 1970). Beans harvested in June start their development in the dry season and are smaller, have a higher shell percentage and a lower butterfat than the main crop beans harvested from September onwards, which develop during the rainy season. These differences are illustrated by the figures overleaf.

Apart from seasonal changes in rainfall, bean size from the same planting material will vary from one country to another. F_3 Amazon

Month of harvest	June	September
Average bean weight (oven dried) (gm)	0·81	1·18
Average shell (%)	16·3	11·0
Butter fat at mean bean weight (%)	48·0	56·9

cocoa from Ghana has an average bean weight of 1·1 grams, but this figure is 1·2 grams for the same cocoa grown in West Cameroon. This difference must be due to rainfall and possibly other climatic factors which influence pod development.

The variety also influences bean size. The best known case is that of the selections Scavina 6 and 12 selected for witches' broom resistance, which produce beans appreciably smaller than typical Trinidad cocoa when these selections were first propagated. The following figures illustrate the difference between SCA 6 and three Trinidad selections.

Range of dry bean weight (based on weight of 12 beans)

		gm
	ICS 40	18·3–27·0
	ICS 1	14·1–22·7
	ICS 95	12·5–20·7
	SCA 6	7·9–13·0

SOURCE: Bartley (1965).

The precise effect of variety, growing conditions and method of preparation on shell percentage has not been elucidated. There is some circumstantial evidence that shell percentage varies according to method of fermentation. In a cocoa quality survey conducted in Nigeria, samples of beans sold by farmers, and therefore prepared by their methods, were compared with samples of beans collected from farmers' plots and fermented by standard methods (Owolabi, 1972). Beans prepared by farmers' methods had 12·2 per cent shell, while standard methods produced beans with 15·5 per cent shell.

Further evidence is contained in the following figures resulting from fermentation trials conducted in Sri Lanka (Knapp, 1926).

It might be assumed that washing should be resorted to where high

Hours of fermentation	Shell percentage, unwashed beans
42½	10·1
64	11·4
88	12·1

shell percentage occurs, but it is very uncertain whether the expense of washing and the loss in weight would be repaid by any increase in price, as the cocoa market is not geared to differences in shell and fat percentages.

International Cocoa Standards

All cocoa is sampled and inspected at some point between producer and manufacturer, and most cocoa-growing countries have regulations concerning the quality of cocoa exported, together with an inspection service to impose them. In the past these tests have been formulated to meet the circumstances of the individual country. In addition the main cocoa markets have defined certain quality standards. Recently, International Cocoa Standards have been formulated at a series of meetings between producers and consumers under the aegis of the Food and Agriculture Organisation. These standards comprise a model ordinance with definitions and grade standards, and a code of practice detailing methods of sampling.

The model ordinance defines 'Cocoa of merchantable quality' as follows:

(*a*) Cocoa of merchantable quality must be fermented, thoroughly dry, free from smoky beans, free from abnormal or foreign odours and free from any evidence of adulteration.
(*b*) It must be reasonably free from living insects.
(*c*) It must be reasonably uniform in size, reasonably free from broken beans, fragments and pieces of shell, and be virtually free from foreign matter.

Grade standards

Grade	Maximum percentage by count		
	Mouldy	*Slaty*	*Insect damaged germinated flat*
Grade I	3	3	3
Grade II	4	8	6

The number of defective beans is determined by the cut test. When a bean has more than one defect, it is recorded in the most objectionable category, the three categories above being in descending order of gravity from left to right.

The full text of the Model Ordinance and Code of Practice is given in Appendix 1.

The cut test

The cut test is the accepted method of assessing quality as defined in grade standards, and by revealing the colour of dried cotyledon or nib, it is also a guide to degree of fermentation. It is therefore a rough guide as to the flavour potential of the sample. The test involves cutting lengthwise 300 beans taken from a random sample of the cocoa whose quality is to be assessed. The recommended method of taking the sample is detailed in the Code of Practice. After cutting the beans, any defective ones are counted, the precise definition of the defect being given in the Model Ordinance. After counting the defects, they are expressed as a percentage.

Mouldy beans

Mouldy beans are defined as beans 'on the internal parts of which mould is visible to the naked eye'. This is the most serious defect of cocoa beans and is objectionable because the presence of mouldy beans affects the flavour of finished chocolate; mouldy off-flavours are unaffected by the manufacturing processes.

Samples of beans with as little as 4 per cent of mouldy beans can produce off-flavours, but there is not necessarily any close connection between percentage of mouldy beans and off-flavours. This may be due to moulds differing in their effect on flavour and also to the fact that mould attack on beans can be invisible to the naked eye. In addition, moulds inside the bean can increase the free fatty acid content of the cocoa butter; this is normally very low—about 1 per cent—but moulds can increase the figure to 4 per cent. There is also the possibility of some moulds giving rise to the presence of mycotoxins; such toxins have only rarely been found in cocoa beans, but toxin-producing strains of some of the fungi found in cocoa beans are known to exist.

A large number of mould species have been found in cocoa beans; in Nigeria, for instance, twenty-eight species have been identified but only thirteen of these occur frequently (Oyeniran, 1973). These moulds invade the beans before harvest, during fermentation or drying, and during storage.

Before harvest, pod disease fungi, *Phytophthora palmivora* and *Botryodiplodia theobromae*, attack the beans, but the latter fungus rapidly overwhelms and replaces the former. Early work by Bunting in Ghana did not find that *B. theobromae* was an important cause of internal mould, but recent work in Nigeria has shown it to be more important there (Broadbent and Oyeniran, 1968). This fungus causes the beans to turn black or brown externally. Black beans can be spotted and rejected as specified in Nigerian regulations, but the black colour can be rubbed off making such beans indistinguishable from healthy ones. Pods attacked by this fungus produce a high percentage of beans with internal mould, but the mould is only visible on microscopic examination in a large proportion of such beans. This fungus is normally inactivated by the heat developed during fermentation and drying, so that it will not develop again later.

During fermentation a few species of fungi are capable of penetrating the testa at a point near the embryo, provided the temperature conditions are suitable. The heat within a fermenting mass is normally too great and it is only round the edges of a mass of beans which is not mixed regularly or is inadequately insulated that such mould attack occurs.

Various species can invade the beans during drying which is prolonged due to dull weather, or some similar cause. The development of internal mould is preceded by external mould growth so that steps should be taken to hasten the drying process when external mould appears.

Finally, moulds can develop during storage. Dried beans will inevitably bear fungal spores and these will develop and may enter the beans if conditions in the bean allow. This can occur if beans are stored in a very humid atmosphere, they will absorb moisture so that the moisture content rises above the critical level of 8 per cent.

Slaty beans

Slaty beans are defined as having a 'slaty colour on half or more of the surface exposed by a cut made lengthwise through the centre'. The slaty colour is distinctive and is readily distinguished from the other colours in the cotyledon both before and after fermentation.

A slaty bean results from drying a bean before any of the initial changes associated with fermentation takes place and they have therefore been killed by drying instead of by heat and acid. Such beans have none of the precursors of chocolate flavour and chocolate made from them has a bitter, astringent and thoroughly unpleasant flavour. As with mouldy beans, properly conducted fermentation will not give rise to any slaty beans.

Infested beans

Cocoa beans are attacked by several stored product pests, though fortunately, dried beans are an unattractive food for most insects. The major stored product pest of cocoa is the tropical warehouse moth, *Cadra cautella*. This moth can infest cocoa as it dries on drying mats or in store. In the tropics it multiplies rapidly, having a life cycle of six weeks; in temperate climates its life cycle is much longer but it can survive a winter in stored cocoa. The major hazard that these moths pose is to other foodstuffs and to the chocolate factory, in which it is relatively easy for them to find conditions suitable for breeding. Moths in a chocolate factory form a serious menace to finished goods.

Other pests, *Araecerus fasciculatus* (the coffee bean weevil), *Lasioderma serricorne,* the tobacco beetle, and *Tribolium castaneum,* the rust-red flour beetle, will attack cocoa in West Africa. The usual source of infestation is other food crops on which the insects have a shorter life cycle than on cocoa. None of these pests will breed in temperate climates, so they do not present a threat to chocolate factories.

Infestation can be avoided or prevented without much difficulty by maintaining clean stores and by ensuring that infested produce is not stored near cocoa. Where infestation is present it can be controlled by fumigation with methyl bromide or phosphine; the latter is easily applied by means of tablets of aluminium phosphide, which break down in a humid atmosphere, giving off phosphine gas.

Other defects

Germinated beans are considered a defect because the hole in the testa left by the emerging radicle offers an opening for moths and insects to invade the beans.

Flat beans are those which have no cotyledon and are therefore useless.

A further defect mentioned in the definition of cocoa of merchantable quality is smoky bean. Such beans have a smoky smell or taste, due to contamination with smoke during drying or storage. The smoky taste is not removed by normal factory processing, so this defect is a serious one. At one time, cocoa from West Cameroon had this fault, and suffered a heavy discount, but better methods of drying have eliminated the defect. Where wood-fired dryers are used, care must be exercised to ensure that smoke does not escape from the flue.

White spot

This term is applied to an unusual condition of dry cocoa beans in which some white spots are seen on the internal surface of the cotyledons and shell. These spots could be confused with moulds, but they have been found to consist largely of theobromine (Benize *et al.,* 1972). The presence of white spot is quite harmless and is not thought to affect flavour. This condition is associated with the harmattan in West Africa or any severe dry season, but the precise reason for the development of white spot is not known.

Assessment of fermentation

In order to assess the degree of fermentation of the beans they can be divided into four categories: fully fermented; partly brown, partly purple; fully purple; and slaty.

Slaty beans are unfermented and a defect dealt with in grading regulations. The other three categories have been defined as follows (Cocoa, Chocolate and Confectionery Alliance, 1968):

> The first category (fully fermented) should include all fully fermented beans, even though the colour cannot properly be described as brown.
>
> The second category (partly brown and partly purple) should include all beans showing any blue, purple or violet colour on the exposed surface, whether suffused or as a patch.
>
> The third category (fully purple) should include all beans showing a completely blue, purple or violet colour over the whole exposed surface. It should also include, irrespective of colour, any beans which are slaty but not predominantly so. ['Predominantly' in this context means more than half.]

The colours of a normal sample of cut beans cover a range from the chocolate brown of fully fermented beans to the fully purple beans that have been inadequately fermented. While the definitions of the three categories are a guide, the differences cannot be defined precisely and the assessment according to these categories is to some extent subjective. Fully purple beans which have the same cheesy texture as slaty beans are rarely present in any sample in excess of 2 or 3 per cent and will be absent from samples prepared by normal large-scale methods. Assessment into the other two categories is imprecise, but it is the relative proportion of these two categories that gives a guide to the degree of fermentation. It is not possible to prepare a sample with 100 per cent fully fermented, and it is not desirable to attempt to do so. Beans described as 'partly brown—partly purple' are not defective and should be present at least to the extent of 20 per cent. A proportion of 30 to 40 per cent is acceptable, but samples with more than 50 per cent in this category have probably been inadequately fermented for some reason and may give rise to bitter and astringent flavours.

The purple colour is due to the presence of unchanged anthocyanins in the dried beans. This constituent of the bean is hydrolysed during fermentation and changed to a colourless leuco-anthocyanin. A change in flavour is associated, directly or indirectly, with this change in colour and beans with 30 per cent or more of unchanged anthocyanin have a deep purple colour and chocolate made from them would have a harsh and bitter taste (Rohan, 1963). The colour of cut beans is usually brighter and more purple soon after drying is completed than some months later when samples are examined in temperate climes. There is a gradual change in colour with storage, published data indicating that samples containing 50 to 70 per cent purple beans initially, frequently halved this proportion after six months storage (Wickens, 1954). This change is associated with a decrease in anthocyanin content; it has been shown that as much as 50 per cent of the anthocyanins can be lost over a four to five month period of storage in Ghana (Kenten, 1965).

Over-fermentation can be revealed by a dull dark appearance of the beans when cut but, like purple beans, such beans cannot be defined. When beans are over-fermented, unpleasant smells develop in the fermenting mass. These arise from breakdown of protein and production of ammonia. This inevitably leads to loss of chocolate flavour and production of unpleasant off-flavours.

References

Bartley, B. G. D. (1965) 'Seed size inheritance', *Ann. Rep. Cacao Res. 1964*, Trinidad, pp. 32–4.

Benize, M., Hahn, D. and Vincent, J. C. (1972) 'Le "White spot" des fèves de cacao', *Café, Cacao, Thé*, **16**, 236–42.

Broadbent, J. A. and Oyeniran, J. O. (1968) 'A new look at mouldy cocoa', *Proc. 1st Internat. Biodeterioration Symp.*, 1968, pp. 693–702.

Cocoa, Chocolate and Confectionery Alliance (1968) *Raw Cocoa: Manufacturers' quality requirements*, 2nd edn, London, CCCA.

Kenten, R. H. (1965) 'Loss of anthocyanins during the storage of dry fermented cocoa', *Ghana J. Sci.*, **5**, 240–2.

Knapp, A. W. (1926) Letter to Dir. Agric. Ceylon, dated 17 February 1926.

Owolabi, C. A. (1972) 'Cocoa quality survey', *Ann. Rep. Cocoa Res. Inst. Nigeria 1969–70*, pp. 33–4.

Oyeniran, J. O. (1973) 'Internal mouldiness of cocoa; causes and control', *Rep. Training Seminar on Cocoa Grading, Lagos 1972*, FAO No. TA 3195, pp. 138–46.

Rohan, T. A. (1963) *Processing of raw cocoa for the market*, FAO Agric. Studies No. 5, p. 96.

Toxopeus, H. and Wessel, M. (1970) 'Studies in pod and bean values of *Theobroma cacao* in Nigeria. I. Environmental effects on West African Amelonado with particular attention to annual rainfall distribution', *Neth. J. Agric. Sci.*, **18**, 132–9.

Wickens, R. (1954) 'Cacao fermentation and quality', *Rep. W. Afr. Cocoa Res. Inst. 1953–54*, pp. 41–3.

Wood, G. A. R. (1973) 'The importance of quality and grading from the producers' point of view', *Rep. Training Seminar on Cocoa Grading, Lagos 1972*, FAO No. TA 3195, pp. 59–65.

Chapter 14

Marketing

A. P. Williamson

Cocoa marketing is the process whereby the ownership of cocoa beans is transferred from the grower to the manufacturer. In the days when cocoa and chocolate were still luxury articles producers sent their cocoa to agents in London and other consuming centres where it was sold by auction or by negotiation. Manufacturers received samples, which they roasted and tested, and then they bid for the parcels which suited their requirements. Auctions in their original form were known as 'sales by candle' from the practice of having a lighted candle, with pins stuck at intervals down its side, set on the auctioneer's desk. The last bid before the pin fell out secured the cocoa.

Since the turn of the century production and consumption of cocoa has increased from 100,000 tons to about 1,500,000 tons a year. To ensure smooth evacuation from the farms, organised shipment to consuming countries and regular supplies to the manufacturers a complex and sophisticated marketing system has evolved.

World cocoa production can be split into two kinds, bulk cocoas which account for 90 to 95 per cent of the total, and fine or flavour cocoas.

Flavour cocoas

Fine or flavour cocoas consist of beans which have distinctive flavours and come largely from Criollo and Trinitario trees. They are used by manufacturers for blending with bulk cocoas and in general they are only available in small and highly individual parcels. The way they are marketed has not changed significantly over the years. They are nearly all bought on sample and individual plantations or shippers appoint agents in the consuming countries to handle their sales, and great care is taken to guard their reputations. They normally fetch a premium over the comparable price of bulk cocoas and this premium may range from a few pounds to over £100 a ton for the top marks which are only available in very small quantities. These premium cocoas can be considered as a separate and high specialised market which forms a declining proportion of the cocoa market as a whole.

Bulk cocoas

Bulk cocoas come principally from the Forastero varieties grown in West Africa and Brazil and they account for an overwhelming proportion of today's world cocoa production. The industry in these countries is almost entirely in the hands of smallholders and each individual grower's cocoa is bought by commercial or government concerns who arrange for the bagging, grading and ultimate sale of the crop.

In Ghana and Nigeria trading companies and some manufacturers set up buying agencies between the wars to buy cocoa from the farmers and sell it on world markets. After the war Marketing Boards were set up and given sole responsibility for the collection and sale of all cocoa production. The principal object of these Marketing Boards is to insulate the farmer from the wide fluctuations in world market prices and to ensure that their cocoa is of a good and uniform quality. Towards this end they fix the price the farmer is paid for his cocoa each season and lay down minimum grading standards. They appoint licensed buying agents, who compete in the collection of the cocoa through stores sited in all the main towns and villages and arrange for the movement of cocoa to port or to marketing board stores. In return the licensed buying agents are paid a buying commission per ton of cocoa purchased to cover expenses and to provide them with a reasonable profit. The Marketing Boards then arrange for the sale of this cocoa on world markets through subsidiary Marketing Companies which were originally situated in London but are now in Accra and Lagos.

In the Ivory Coast and Cameroon individual shippers have intermediaries who have buying centres throughout the cocoa areas. Originally there was no control over the price they offered the farmer and it used to vary daily in line with prices on the world market. In 1956–57 however, the French Government set up a *Caisse de Stabilisation* in each country. These have similar objects to the Marketing Boards in Ghana and Nigeria but leaving things more in commercial hands. A minimum producer price is fixed for each season and the *Caisse* also declares each day a nominal world price based on quotations on the New York, Paris and Amsterdam markets. Shippers continue to operate as before but either receive from or pay to the *Caisse* the difference between the nominal price for the day and the producer price after deduction of expenses at an agreed rate. More recently the *Caisses* have started to sell themselves and they have also blocked sales by exporters when they do not think the market situation is right.

In Brazil, the remaining major producer of bulk cocoa, production is on plantations as well as on smallholdings. Selling arrangements are governed by the foreign trade department of the Bank of Brazil, known by its Portuguese abbreviation CACEX. This body indirectly controls when the shippers are selling, although the price they pay the farmer varies from day to day in line with world prices.

Actuals market

The selling of most of the world's cocoa is thus in the hands of a limited number of marketing companies, or shippers under some central control. These have become known as the origin markets and the principal centres are in Accra, Lagos, Abidjan, Douala and Salvador. Each origin sells a standard type of cocoa and they invariably sell forward, i.e. for shipment during some specified period in the future.

Anyone can approach the shippers or marketing companies to buy cocoa, although there is usually a minimum quantity they are prepared to sell. In the case of the Ghanaian and Nigerian marketing companies it is 200 tons. Buyers of cocoa from origin can be divided into two kinds: manufacturers who want the cocoa for their own use, and dealers who buy the cocoa to resell it or to fulfil a sale they have already made.

It might be useful at this point to distinguish between a broker and a dealer. A dealer buys cocoa on his own account and may or may not have a prospective customer in mind. He buys and sells cocoa which is either in store, on board a ship or awaiting shipment and may sell cocoa before he has actually bought it. The broker on the other hand merely brings together a buyer and a seller and although he may become principal to both contracts he conducts both the buying and selling side simultaneously. Normally he will be advising one or both parties and will arrange the contract details, and for this he is paid a commission or brokerage, which in London is half per cent of the contract value. Sellers and buyers employ brokers for a variety of reasons. For instance, they may only deal occasionally in cocoa and require the broker's expert knowledge, or they may be large suppliers, users or even dealers who want to deal through an intermediary to keep their own identity a secret. Increasingly the distinction between brokers and dealers is being broken down and there are now very few brokers who are not dealers as well.

Dealers who buy cocoa to resell it to manufacturers and other dealers and are prepared to sell cocoa they have not yet bought, form the basis of the 'secondhand' market which is used by most manufacturers because they are too small to justify the expense of dealing direct with the origins. Even the larger manufacturers often find it more convenient to use the secondhand market as origins may be withdrawn from the market or may not be offering the shipping period required.

In addition a dealer can protect both the manufacturer and the producer from the commercial risks involved in trading around the world. The principal secondhand markets are in London, New York, Amsterdam, Hamburg and Paris; London is by far the largest.

The above are all methods of buying and selling physical cocoa and for that reason they are termed the 'actuals' or physical cocoa market.

Most sales of cocoa are made on standard contract forms, drawn up and controlled by one of a small number of trade associations that have developed as the trade in cocoa expanded. They also regulate disputes and generally look after the interests of all those involved. The Cocoa Association of London, which consists of dealers, brokers, manufacturers, producers and shippers, was formed in 1926. In the United States there is the Cocoa Merchants Association and in France the 'Association Française du Commerce des Cacaos'. French Association contracts are normally used for cocoa from the ex-French territories, the London Association contracts for cocoas from the ex-British territories and the American Association contracts for cocoa shipped to the United States. One or two of the origins now use their own contracts but they follow the Associations' contracts very closely and in the event of dispute they use the arbitration procedures of one of the Associations.

Sales are usually made on an fob, cif or 'ex store' basis. Fob (free on board) means that the buyer accepts responsibility for all charges from the point when the cocoa is loaded over the ship's rail in the producing country. In a cif contract (cost, insurance and freight) the seller is responsible for all charges up to the point where the cocoa is delivered over the ship's rail at the port of discharge. In both cases the buyer pays on receipt of the bill of lading, which is his entitlement to collect the goods, but in the case of a cif contract the bill of lading is endorsed freight paid whereas in the case of an fob contract the buyer has to pay the freight before the shipper will release the cocoa. Ex store, as the name implies, means the sale of cocoa in a warehouse. Such sales are normally confined to the secondhand market in consuming countries and the cocoa may be immediately available (spot cocoa) or available at some specified time in the future.

Cocoa from all the major producers is shipped in hessian bags containing 140 lb (63·5 kg) of beans in Ghana and Nigeria, 67 kg in Ivory Coast and Cameroon, and 60 kg in Brazil. The bags are marked to indicate the country of origin, the grade and whether the cocoa is main crop or light, mid or summer crop. Ghana and Nigeria cocoas are invariably sold on the basis of the nominal bag weight (shipping weight) with the buyer having a claim if the weight is not within 1·5 per cent of the nominal weight. Ivory Coast and Cameroon cocoas are normally sold on the basis of actual landed weight, as are most ex store contracts.

In the case of cif and fob cocoa the seller is usually allowed to ship

1. Botanical illustration from *Der Cacao und die Chocolade* by Alfred Mitscherlich, Berlin 1859. The original lithograph was by Carl Friedrich Schmidt.

1 The parts indicated in the main part of the illustration will be obvious.
3 (*a*) sepals   (*b*) petals
6 *(a)* staminodes   (*b*) stamens   (*c*) anthers   (*d*) ovary

II. Symptoms of nitrogen deficiency. Three leaves showing different degrees of deficiency. The older leaf on the left shows tip scorch to a small degree.

III. Symptoms of iron deficiency.

IV. Symptoms of potassium deficiency.

V. Symptoms of calcium deficiency.

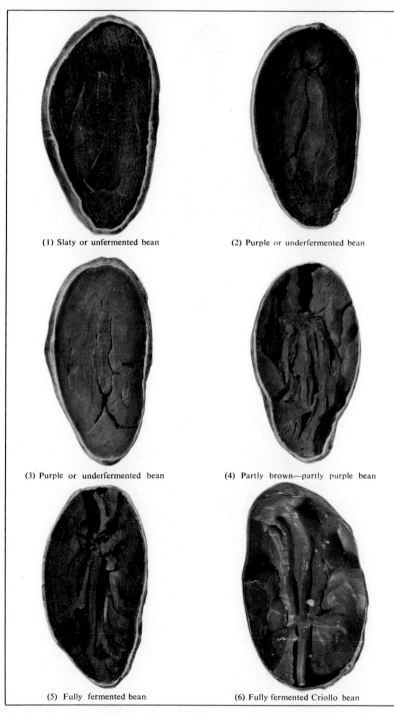

(1) Slaty or unfermented bean

(2) Purple or underfermented bean

(3) Purple or underfermented bean

(4) Partly brown—partly purple bean

(5) Fully fermented bean

(6) Fully fermented Criollo bean

VI. Sections of cocoa beans: Nos. 1-5 Forastero beans.

the cocoa at any time during a specified two or three months period. During loading or unloading, depending upon whether it is an fob or cif contract, the cocoa is weighed and sampled by independent supervisors appointed by buyers and sellers, who check the weights and see that samples are properly taken or 'drawn'. Should the buyer consider that the samples are not up to the contract standard or, if the cocoa has been bought on the basis of shipping weights that the weight is short, he may claim for an allowance from the seller. He cannot however reject the cocoa. If the seller does not agree and the matter cannot be settled amicably, arbitrators are called in to settle the claim.

The value and price of cocoa

The value of a particular cocoa is dependent on its quality, as discussed in Chapter 13. A vital factor is the amount of cocoa butter that can be extracted. The cocoa butter in the bean is almost entirely contained in the nib, so the amount of butter in the bean depends on the percentage of nib and the percentage of butter in the nib. Ghana cocoa has slightly more nib than Nigerian cocoa and the nib contains a slightly higher percentage of cocoa butter. Manufacturers therefore get a little more cocoa butter from a given tonnage of Ghana beans than from a similar tonnage of Nigerian beans. In addition, although both Ghana and Nigeria ship in 140 lb bags, in practice Ghana cocoa is nearer the nominal 140 lb a bag than Nigerian cocoa, although both are within the contract limit of 1·5 per cent. Thus on the grounds of its higher fat content and better bag weights, Ghana cocoa normally fetches a small premium over Nigerian cocoa.

New growers of bulk cocoas, particularly those from new producing countries, have to sell their cocoa on sample through brokers or dealers in a similar manner to the way in which fine and flavour cocoas are sold as the quality and flavour of their cocoa is not known to the market and manufacturers. Small parcels of bulk cocoa in this category are not easily sold and until the producer has established a reputation he may find it necessary to accept a discount to the prices being paid for the main bulk grades.

The price of a particular type of cocoa in relation to other cocoas on the market at any given time does not depend solely on its economic value. It is also dependent on the demand, or lack of demand, for the cocoa in question. If Ghana cocoa is in comparatively plentiful supply when Nigerian cocoa is scarce, then Nigerian cocoa can sell at parity to Ghana despite its lower intrinsic value.

The price of cocoa is also affected by the period of shipment. Normally one has to pay more the further ahead the shipment period. This is because in times of adequate supply dealers and manufacturers have to be induced to carry the surplus cocoa and they will only do this if they can buy the cocoa and pay the costs of storage and finance for less than they would have to pay if they bought the cocoa for later shipment. In this situation one is said to have a 'carry' in the market. In times of shortage, however, stocks get used up and the demand is for cocoa for immediate delivery, not for cocoa for delivery in the future. In this situation one gets what is called an 'inverted' market with the immediate or 'nearby' deliveries at a premium to the later deliveries.

Finally, of course, the price of cocoa is also affected by the port to which it is to be shipped. The normal basis is either 'main north European port' or in the case of America 'Atlantic and Gulf port'. Cocoa to all other ports will be at a premium or 'freight differential' to the rates for these base ports.

Terminal market

The other market that has to be considered is the terminal or 'futures' market. While the great majority of the trade in cocoa in both the actuals and terminal markets involves future shipment or delivery, the terminal market is rather different in that the normal practice is for contracts to be bought back or resold before the stipulated delivery period, with a cash settlement taking place to cover any changes in price rather than the buyer actually taking delivery from the seller. As a result it is often known as the 'paper' market and to appreciate how it works it is necessary to examine it in more detail.

The two principal terminal markets, the London Cocoa Terminal Market and the New York Cocoa Exchange, were formed in the 1920s to provide a regular floor for buying and selling cocoa. More recently markets have been opened in Amsterdam and Paris. A London terminal contract stipulates the quantity, in multiples of 10 ton lots, of a standard grade of cocoa, the month of delivery and the price. For convenience the months for delivery, or 'positions' as they are termed, are limited to March, May, July, September and December. At any given time cocoa can be traded for delivery in these positions for up to eighteen months ahead. While a standard cocoa is specified in the contract the seller may deliver any of a wide number of types of cocoa stipulated in the market rules, with an adjustment being made to the price by standard discounts or premiums depending on the cocoa actually tendered. The seller also has the right to tender the cocoa in store in London, Liverpool, Avonmouth, Hull or Middlesbrough in the United Kingdom, and in Amsterdam, Antwerp, Hamburg or Rotterdam on the Continent. Thus the buyer of a terminal contract does not know what type of cocoa he is going to get nor where he is going to get it, and this is one of the main reasons why delivery is rarely taken of terminal contracts.

There are fifty-four full members of the London Cocoa Terminal

Market Association, and they alone can trade on the floor. Everyone else, including associate and overseas members of the market, have to put their business through one of the trading members. When a trading member is instructed by a client to buy, say one 10-ton lot for December 1973 delivery at the market price, he bids in the ring and when another floor trader accepts his bid a binding contract is made. Each day many thousands of lots change hands and at the end of trading they all have to be confirmed. To do this each trading member confirms its own position with an independent body, the International Commodities Clearing House, which has been appointed by the market to safeguard the terminal market contract. Once this body has confirmed the contract with both the buyer and the seller it guarantees the fulfilment of the contract to both parties and in effect becomes the other party to every transaction. Thus if trading member A buys one lot of December 1973 delivery from member B and subsequently sells one lot to member C he can 'liquidate' his position by putting the purchase against the sale and receive from or pay to the Clearing House the difference in the contract prices. This will leave member B who sold him the cocoa still 'short' of one lot and member C who bought the cocoa from him still 'long' of one lot.

In the above example, having bought one lot of cocoa the member A liquidated the contract by selling one lot. As we have seen, because of the uncertainty about the type and location of the cocoa that will be tendered this is normally the way in which a terminal market transaction is completed, i.e. one makes or receives a cash settlement to offset any change in the market price between the original purchase price and the final sale price.

A member who has sold cocoa may of course deliver or 'tender' the cocoa he has sold; similarly, a buyer may hold on to a contract and take delivery. This is the essence of the terminal market as it ensures that terminal prices move in line with actuals prices. If for some reason terminal prices are cheap in relation to actuals prices then someone who has bought terminal cocoa will find it advantageous to hold on to the contract and take delivery. Conversely if the terminal prices are high in relation to actuals prices then it will pay a dealer or manufacturer to buy cocoa 'in store' and to simultaneously sell the first or 'spot' terminal delivery position to tender the cocoa.

The value of the terminal market

A terminal market that moves up or down in line with the actuals market provides everyone involved in trading cocoa, whether producer, consumer or dealer, with an opportunity of insuring or 'hedging' against the risk of unwanted price changes by selling on the terminal market against the purchase of actual cocoa or by buying on the terminal market against the sale of actual cocoa.

For example a dealer may contract in January to supply a manufacturer with July/September shipment Ghana cocoa cif UK at a premium of £7·50 a ton over the price of the September terminal position. If the dealer is unable to buy the cocoa from origin or from another dealer on the secondhand market and does not want to speculate that the price will fall before he can buy the cocoa he will buy a similar quantity of cocoa on the terminal market. His position might then be:

Sold	July/September	Ghana cif UK	@	£307·50/ton
Bought	September terminal		@	£300·00/ton

The actuals and the terminal markets move up and down together so that if the dealer ultimately has to pay more for the July–September shipment cocoa to fulfil his contract, then his terminal contract will have risen by a similar amount and the loss on the actuals transaction will be offset by a profit on the terminal deal. On the other hand if prices fall and he makes a loss on the terminal operation, this will be offset by the profit he makes by being able to buy the actual shipment cocoa he requires at a lower price.

He is thus hedged against any movement in price and his profit will not depend on the movement in prices but on his skill in relating his offer price to the going terminal price so as to leave himself sufficient margin for profit. In the example being considered, three months later in April prices may have risen by £50 a ton but he is then able to buy the cocoa he requires at a premium of £5 a ton to the price at which he can sell the September terminal contracts so his position would become:

January—*Sold*	July/September	Ghana cif UK	@	£307·50/ton
April —*Bought*	July/September	Ghana cif UK	@	£355·00/ton
				£47·50/ton loss
January—*Bought*	September terminal		@	£300·00/ton
April —*Sold*	September terminal		@	£350·00/ton
				£50·00/ton profit

Net profit .. £2·50/ton

Thus the dealer estimated correctly that by waiting until the origins were selling July/September shipment cocoa he would be able to buy from origin or in the secondhand market at £5 a ton above the price of September terminal, whereas he had been unable to obtain the cocoa at £7·50 a ton premium when he originally sold the cocoa.

The terminal market also enables producers to be much more flexible in selling their cocoa. For instance when prices are high an estate might decide that, at the price levels ruling at the time, it can make a very satisfactory profit and be prepared to forego the possibility of additional profit if prices rise further so long as it can be assured of this profit should prices fall. At the time it might not be possible for the estate to find a suitable buyer for its production, or it might not want to commit itself for a specific tonnage at a specific time so far in advance of the crop. In this situation it can sell an equivalent quantity of terminal cocoa for delivery say in twelve months' time. If the market falls the estate will get a lower price when it sells its production but this will be offset by the profit it can make by buying back its terminal sale. Conversely, if prices rise the estate will get a higher price when it sells its cocoa but there will be a loss when it has to buy back its terminal sale. Either way the estate would receive the sale price on which it had budgeted when it made the terminal sale.

So far we have looked at ways in which the terminal market can be used by dealers, manufacturers and producers to hedge themselves against unwanted price movements while they get on with their main business of producing, distributing or processing cocoa. At the same time there are people who are prepared to put their money at risk in the hope of making a profit. These speculators or 'risk takers' may be producers, dealers or manufacturers trying to beat the market or they may be people outside the cocoa industry. They are an essential element of the cocoa market as without them there would not be a regular trading market in which to hedge purchases and sales. At times, however, particularly in times of shortage, speculators can accentuate market movements although they cannot create the movements.

A speculative use of the terminal market can be of great value to an estate in helping to free it from the restrictions placed on it by the seasonal nature of its crop. For instance if, when prices are low, an estate had its store full it might be obliged to sell its cocoa even though it felt that prices were likely to rise. In this instance it could proceed with the sale of the actual cocoa but at the same time it could buy back a similar quantity of cocoa on the terminal market. If prices do subsequently rise then there will be a profit when the terminal purchases are sold to be added to the price obtained for the cocoa. If prices fall further there will of course be a loss which would have to be deducted from the price received from the sale of the cocoa. The point is that by making an offsetting terminal purchase when it sold the cocoa, the estate was able to avoid having to commit itself on price just because its store was full.

International Cocoa Agreement

It would be wrong to conclude even such a short introduction to the marketing of cocoa without reference to the International Cocoa

Agreement which came into effect on 1 October 1973 after more than sixteen years of negotiation.

The object of the agreement is to reduce the wide fluctuations in the price of cocoa and to establish a minimum price. This is to be done by restricting cocoa exports as prices fall, the surplus being held in a buffer stock to be sold when prices rise again. In this way it is hoped to control prices in an agreed range, which for the first two years will be from 23 to 32 cents a pound. It should be recognised, however, that while there is a mechanism to prevent the price of cocoa falling below the minimum, there is nothing to prevent the price rising above the top end of the price range once any cocoa in the buffer stock has been sold. The top end of the range is not therefore a maximum price.

All the major producers and consumers except the United States are members of the agreement, which is controlled through a Cocoa Council on which all members are represented. Prior to the start of each crop year, which runs from 1 October to 30 September, the Cocoa Council decides on the level of total world exports required to keep prices around the middle of the range based on the forecast level of demand. The exports from countries producing less than 10,000 tons and certain listed producers of fine and flavour cocoa are deducted as there are no restrictions on their exports; and in addition an allowance is made for exports from non-member countries. The balance is divided among the major producing countries in proportion to their basic quotas to give each an export quota for the year. The basic quotas are based on each country's highest production in recent years.

These annual export quotas are adjusted up or down according to the indicator price ruling at the time on the following basis:

cent/lb	%
below 24	90
24–26	95
26–27·5	100
27·5–29	105
above 29	no restrictions on exports.

If a country produces more cocoa than its final quota for the year, it will be able to sell the surplus to the buffer stock. The buffer stock will initially pay 10 cents a pound fob, but when the cocoa is ultimately sold the producing country will receive the sale price less the 10 cents a pound and the costs incurred in shipping and storing it. The buffer stock is limited to 250,000 tons and once this tonnage is reached any further cocoa will be bought by the buffer stock and sold to vegetable oil processors. If the indicator price reaches 31 cents a pound the buffer stock manager may commence selling, and at 32 cents a pound he must sell any cocoa remaining in the buffer stock. The buffer stock operation is financed by a levy of 1 cent a pound on all cocoa exported by the member producing countries except producers of fine and flavour cocoas. All member producing countries require that a certificate of contribution showing that a levy has been paid be produced before allowing cocoa to be exported and all importing members require that the certificates be produced at the time of import.

Chapter 15

Production and Consumption

The development of the cocoa industry to 1940 has been reviewed in Chapter 1; the changes that have taken place since the war are described here.

Production

At the end of the war cocoa production was around 600,000 tons, which was 100,000 tons lower than in the five years prior to 1940. Production recovered to its prewar level by 1950 but remained relatively static at 700,000 to 800,000 tons during most of the 1950s. In the crop year 1958–59 production started to move up, and rose to 1,173,000 tons in 1960–61. Production remained at that level until 1964–65, when there occurred a remarkable increase of over 20 per cent to 1,484,000 tons. In that year there was a bumper crop in Ghana, Nigeria and Ivory Coast, the reasons for which remain obscure, as it cannot be attributed to action common to all three countries. In the following year production declined to its former level, since when it has risen irregularly to about 1·5 m tons.

In 1946–47 the five leading producers were Ghana, Nigeria, Brazil, Ivory Coast and Cameroon in that order, and they provided 77 per cent of total world production; in 1970–71 the same five countries were in the lead, each producing over 100,000 tons, 78 per cent of the world total. This might seem to indicate a relatively stable situation, but this is far from true as there have been many vicissitudes during those twenty-five years which have been reflected in widely fluctuating prices. It is not the purpose of this chapter to describe these developments in detail, but to outline the major changes that have taken place in West Africa and elsewhere.

Details of world production since 1945 are given in Table 15.1.

Ghana

At the beginning of this period the future of cocoa production in Ghana

Table 15.1 *World production of raw cocoa (thousand long tons)*

Country	1945–46	1950–51	1955–56	1960–61	1965–66	1970–71	1975–76
Africa							
Cameroon	34	47	53	73	78	110	94
Gabon and Congo	—	2	3	5	3	7	7
Ghana	209	262	237	433	410	386	391
Guinea (Equatorial)	17	14	19	25	34	32	12
Ivory Coast	28	56	70	93	111	177	227
Nigeria	103	110	114	195	182	303	212
São Tomé	10	8	8	10	8	10	8
Sierra Leone	1	2	2	3	5	5	6
Togo	3	4	6	12	15	28	17
Zaire	1	2	4	6	4	6	5
Other	2	1	1	2	2	20	6
Total Africa	408	508	517	857	852	1,084	985
America							
Bolivia	2	3	3	2	2	1	3
Brazil	138	153	168	122	170	179	255
Colombia	11	14	15	19	17	21	26
Costa Rica	5	3	7	13	7	4	6
Ecuador	17	32	32	41	35	60	61
Mexico	4	8	14	27	24	27	31
Panama	2	2	2	1	1	—	—
Peru	3	5	4	3	2	2	2
Venezuela	18	17	23	17	20	18	15
Other	1	1	2	2	2	2	2
Total America	201	238	270	247	280	314	401
West Indies							
Cuba	2	2	2	3	2	1	1
Dominican Republic	25	31	28	34	28	25	29
Grenada	2	3	1	2	2	3	3
Haiti	1	2	2	2	3	4	3
Jamaica	1	2	2	2	2	2	1
Trinidad and Tobago	4	9	9	6	5	4	3
Other	1	1	1	1	1	—	—
Total West Indies	36	50	45	50	43	39	40
Asia and Oceania							
Indonesia	—	1	1	1	1	2	3
Malaysia	—	—	—		1	4	17
New Guinea	—	—	1	7	18	29	30
New Hebrides	2	1	1	1	1	1	1
Philippines	1	1	1	3	4	4	4
Sri Lanka	2	2	3	3	2	2	2
Western Samoa	2	2	4	4	3	2	2
Total Asia and Oceania	7	7	11	19	30	44	59
WORLD TOTAL	652	803	843	1,173	1,205	1,481	1,485

SOURCE: Gill and Duffus Reports.

appeared to be threatened by swollen shoot disease. First reported in 1936 and diagnosed as a virus disease in 1939, it had spread from its initial focus of infection to infect a large proportion of the cocoa in the Eastern Region. There seemed to be little reason to believe that the increasingly important area in Ashanti would not succumb in time.

With promptings from manufacturers, the United Kingdom Government built up a large cocoa survey organisation whose task was to survey the cocoa farms, locate any outbreaks of disease and to control those outbreaks by cutting out and regular inspection. Largely because of the war these measures were taken too late to prevent a vast amount of damage to the cocoa area in the Eastern Region, a part of which was abandoned at an early stage, but they proved effective in preventing serious spread of disease to Ashanti.

The organisation continued its work after independence, but in 1962 a radical change in the control of swollen shoot disease was introduced, the farmers being made responsible for cutting out while the extension staff identified and marked diseased trees. The results were disappointing and at the end of 1964 the Government started to resume the task of cutting out. By that time much of the large organisation involved had been disbanded and it took several years to restore the survey and disease control work, the scale of which can be judged by the fact that over 135 m trees had been cut out between 1947 and 1969.

The cocoa survey conducted in the 1950s showed that there were over 1·6 m hectares (4 m acres) of cocoa, 25 per cent of which had not come into bearing. Most of the young cocoa was in Ashanti and Brong Ahafo in the north-west part of the cocoa area, extensive plantings having been stimulated by the sharp rise in farmers' prices after 1947. These new plantings made little impression on production because a large proportion was badly attacked by capsids; this pest also damaged large areas of mature trees, reducing yield considerably. The development of mistblowers and use of BHC provided effective control of capsids, and the provision of spraying machines and insecticides at subsidised prices enabled the farmers to spray large acreages from 1959 onwards. This was a major factor in the rapid increase in production from 255,000 tons in 1958–59 to 433,000 tons two years later. Since then production has fluctuated about that level, with the notable exception of the 1964–65 crop, which rose to 557,000 tons.

The provision of spraying machines and insecticides to farmers was arranged through the Ministry of Agriculture. In the early years from 1960 sales of insecticide amounted to 200,000 to 300,000 gallons a year, but Ghana's economic difficulties lead to a sharp drop in the use of BHC in 1965. These difficulties have been gradually overcome and more insecticide has become available but strains of capsids resistant to BHC have evolved and have been spreading slowly. An alternative insecticide is available.

The future of Ghana's cocoa production depends on raising the yield in the old areas because there is little scope for new plantings and large areas are old and yielding poorly. The rehabilitation of old areas including the replanting of farms is being tackled, the new planting material developed by the Cocoa Research Institute being used. The major effort in this direction lies in the Eastern Region where a rehabilitation scheme covering 35,000 hectares and financed by World Bank funds was started in 1971.

Nigeria

In the period 1945–50 production in Nigeria was about 100,000 tons, nearly all of which was produced in the Western Region. By 1970 production had risen to 300,000 tons, 95 per cent of which comes from the Western Region.

As in Ghana the threat of virus disease seemed severe in the postwar years, at which time much of the cocoa around Ibadan—the heart of the cocoa area at that time—was infected. This area was sealed off by a cordon sanitaire, and this measure, or the mild nature of the viruses in Nigeria, resulted in limited spread of the disease. Studies of the virus disease showed that, where conditions were suitable for cocoa and with control of capsids, the virus occurring in Nigeria had little effect on yield. The survey and disease control measures were therefore abandoned during the 1960s and currently virus disease does not seem to present a serious threat to the cocoa area.

The 'centre of gravity' in the Western Region, now the Western State, has moved eastwards from Ibadan to Ife and Ondo. The new areas are liable to suffer heavy losses from black pod but control measures using copper fungicides were demonstrated to farmers in 1953 and eagerly taken up by them. A few years later the control of capsids by BHC was similarly demonstrated and adopted by farmers. These three factors—new plantings and control of black pod and capsids—account for the considerable rise in production.

Regional Production Development Boards, founded after the war, established many plantation projects, including some cocoa estates. Cocoa was planted on three estates in the Western Region and on at least seven estates in Eastern Nigeria. In the west yields on estates did not exceed farmers' yields and in the east the estates did not come into production before the civil war lead to a period of neglect.

As in Ghana the area of unplanted land suitable for cocoa has diminished, and rehabilitation and replanting have become necessary for maintenance or improvement of production. A scheme covering 17,000 hectares and financed by the World Bank was started in 1971.

Ivory Coast

The Ivory Coast has shown the most spectacular increase in production of any country in West Africa. After the war production was 30,000 tons, and increased to 180,000–200,000 tons by 1970. At the beginning of this period cocoa was produced in areas adjacent to Ghana, but since then cocoa has spread westwards, and has been promoted in preference to coffee, of which the Ivory Coast was one of the largest producers in Africa.

Increases in production have arisen from better standards of husbandry and control of pests and diseases instigated by the extension service which works through groups of farmers rather than individuals. There are still large areas of land suitable for cocoa, the limiting factor being black pod disease which increases in severity towards the west where the rainfall is heavier. Plans for cocoa, aided by the World Bank, envisage production rising to 340,000 tons by 1980.

Cameroon

There are two cocoa areas in Cameroon, a large one in the east producing 80 to 90 per cent of the cocoa and a small area in West Cameroon, formerly part of Nigeria. In the eastern area cocoa is grown entirely by smallholders while in the west there are one or two cocoa plantations. In both areas black pod is severe, particularly so in the west where the climate is wetter and more humid. Both areas have increased production steadily from a total of 40,000 tons in 1947 to 100,000 tons in 1970. Further increases will rest largely on control of black pod disease in the east, while in the west some new areas could be planted if roads were improved.

Brazil

During the twenty-five years to 1970 production in Brazil has fluctuated widely, but overall has increased gradually from 100,000 tons in 1947 to 180,000 tons in 1970. This represents a fall in Brazil's proportion of the world crop from 17 per cent to 12 per cent.

Brazil's production is concentrated in the State of Bahia. Rainfall is on average adequate and well distributed but is liable to considerable variation and prolonged droughts, which account for some of the fluctuations in production. The other limiting factors are black pod disease and the age of many plantings. These problems are being tackled vigorously by CEPLAC, a large research and extension organisation, founded in 1957. The activity of the extension work has led to large increases in the use of pesticides and fertilisers, and this seems to have been reflected in the higher production level in recent years.

Far East

The postwar period has seen the emergence of two new producing countries, Malaysia and Papua–New Guinea and stagnation or decline in the older producing countries, Sri Lanka and Indonesia. No serious attempt was made to grow cocoa in Malaysia until the 1950s when attention was drawn to the possibilities following surveys by the British Government and manufacturers. Developments followed in Malaya and in Sabah, cocoa being established on a plantation basis in both countries. Towards the end of the 1950s dieback badly affected the young plantings on the east coast of Malaya. This caused a delay in further planting of cocoa throughout West Malaysia but by the mid 1960s it was clear that cocoa could be grown successfully under coconuts and many plantations adopted this practice.

In Sabah development was slower as cocoa had to be planted after clearing the jungle. Production from Malaysia as a whole had reached 5,000 tons by 1971/72 and will increase. Furthermore success in Malaysia may well lead to a revival of interest in Indonesia.

In Papua–New Guinea cocoa expanded after the war by the under-planting of coconuts followed by planting on new land. Production has reached nearly 30,000 tons.

Consumption

World grindings of cocoa beans have increased in response to rising production. The erratic rise in production has inevitably led to surpluses and deficits and hence to wide changes in price.

During this period there have been profound changes in the pattern of grindings and consumption. At the beginning of the period the United States and United Kingdom were the two largest consumers, but the level of grindings in these two countries has not increased; grindings in the United Kingdom have in fact fallen from an abnormally high level during the war, at which time the usual supplies of cocoa butter from the Netherlands were not available. In Western Europe there was a rapid recovery of grindings in Germany and the Netherlands followed by increases which doubled their prewar consumption; in the Netherlands the annual consumption of 100,000 tons of beans is largely re-exported as butter and powder. There have been increases of similar proportions in several smaller consuming countries.

In Eastern Europe and the USSR increases have been spectacular, from 20,000 tons to nearly ten times that level. Consumption in the USSR has risen steadily to over 100,000 tons, but consumption per head is still less than in the United Kingdom and several other countries in Europe.

In the Americas changes have been relatively small, the largest in proportion being in Colombia and Mexico, two countries which consume relatively large quantities of cocoa products per head.

In Asia there was a rapid rise of consumption in Japan to over 30,000

Table 15.2 *World grindings of raw cocoa (thousand long tons)*

Country	1945	1950	1955	1960	1965	1970	1975
Western Europe							
Austria	13	5	7	10	12	13	11
Belgium	9	8	9	14	18	19	19
Eire	3	3	5	6	10	10	7
France	21	64	43	52	62	39	33
Germany (Federal Republic)	—	50	72	113	155	124	137
Italy	—	10	19	28	41	42	29
Netherlands	—	61	56	84	116	113	117
Spain	14	1:	15	21	30	28	33
Switzerland	7	10	9	11	14	17	15
United Kingdom	115	126	102	74	101	81	72
Other	11	25	15	20	25	24	24
Total	193	373	352	433	584	510	497
Eastern Europe and USSR							
Czechoslovakia	4	2	8	12	14	20	20
Germany (Democratic Republic)	—	1	5	12	15	19	22
Hungary	—	—	2	4	9	10	15
Poland	—	2	5	9	15	15	35
USSR	8	13	16	30	72	100	148
Yugoslavia	—	1	2	3	11	11	15
Other	—	—	—	2	9	13	20
Total	12	19	38	74	145	188	275
Africa							
Cameroon	—	—	4	6	14	31	30
Ghana	—	3	—	4	47	41	48
Ivory Coast	—	—	—	—	13	35	52
Nigeria	—	—	—	—	—	23	19
Other	4	4	5	5	7	7	7
Total	4	7	9	15	81	137	156
America							
Brazil	20	27	31	61	55	60	97
Canada	23	17	13	12	17	15	11
Colombia	16	22	25	24	31	34	31
Ecuador	—	3	4	5	8	12	26
Mexico	5	5	9	16	12	21	27
United States	285	265	188	215	287	261	205
Other	23	27	36	40	36	31	33
Total	372	366	306	373	440	434	430
Asia							
Japan	—	—	3	9	29	34	29
Other	2	4	5	9	23	16	32
Total	2	4	8	18	52	50	61
Australasia							
Australia	13	10	8	11	14	13	14
New Zealand	5	2	2	3	4	4	4
Total	18	12	10	14	18	17	18
WORLD TOTAL	601	781	723	927	1,320	1,336	1,437

SOURCE: Gill and Duffus Reports.

tons, but this level has not changed substantially since 1969. The Philippines has also increased its grindings and a chocolate industry has been started in Singapore. Details of world grindings since 1945 are given in Table 15.2.

The most interesting development during this period has been the rise in grindings in producing countries. In 1945 such grindings amounted to 50,000 tons, most of which was ground in Brazil, Colombia and Mexico. Colombia is a special case because local consumption is high and the industry has to import cocoa in addition to the local production of more than 20,000 tons. Mexico consumes all the cocoa that is ground locally but in this case there is a surplus of cocoa beans for export. Brazil, on the other hand, developed a butter-pressing industry during the war, the cocoa butter being largely exported. A butter-pressing industry developed in the other major cocoa producing countries during the 1960s, so that grindings in producing countries have reached a total of nearly 350,000 tons (see Table 15.3).

Table 15.3   *Grindings of raw cocoa in producing countries (thousand long tons)*

Country	1945	1950	1955	1960	1965	1970	1975
Brazil	20	27	31	61	55	60	96
Cameroon	—	—	4	6	14	31	30
Colombia	16	22	25	24	31	34	31
Dominican Republic	—	6	11	15	5	3	5
Ecuador	—	3	4	5	8	12	26
Ghana	—	3	—	4	47	41	48
Ivory Coast	—	—	—	—	13	35	52
Mexico	5	5	9	16	12	21	27
Nigeria	—	—	—	—	—	23	19
Philippines	—	2	2	5	10	7	7
Venezuela	1	2	5	7	8	7	6
Other	10	12	12	11	10	8	15
TOTAL	51	82	103	154	213	282	362

SOURCE: Cocoa Statistics, Gill and Duffus Group, December 1977.

With the exception of Colombia and Mexico very little cocoa is consumed in producing countries, the processing industries exporting cocoa butter and powder in competition with the long-established industry in the Netherlands. The total trade in cocoa butter and powder has grown enormously; in 1970 the volume of this trade was ten times the level twenty-five years earlier.

Byproducts

There are few byproducts from the production of dry beans and those that exist are not widely made. In Brazil a cocoa jelly is made on a small scale from sweatings, and in Mexico sweatings are sometimes allowed to ferment to produce a cocoa wine. There has been some publicity about a cocoa wine in Nigeria but in this case the wine is made from dried beans, which are roasted, ground and mixed with sugar.

The pod husk is rich in potash and has been used occasionally as a source of potash for soap. Otherwise pod husks are not used although some use for them has been investigated. The use of dried cocoa pod meal in the rations for pigs and dairy cows was tried in Costa Rica (de Alba and Basadra, 1952; de Alba *et al.*, 1954). In both trials the use of pod meal up to 50 per cent of the ration gave satisfactory results; meal from rapidly dried pods was equal to good quality maize but pods dried in the sun gave poorer quality meal. It is not known whether the cocoa pod meal can be produced at a price competitive with other feeds.

The use of cocoa byproducts from the manufacturing end—cocoa shell and cocoa meal—for livestock has to be severely restricted owing to the presence of small amounts of theobromine. Some similar precautions might have to be taken in the case of pods as they too contain a little theobromine.

Interest in byproducts has revived and some new uses may be found. The subject of byproducts and alternative uses for cocoa beans was reviewed by Greenwood-Barton (1965) and their use in the feeding of livestock has been described by Owusu-Domfeh (1972).

Research

Research on the cocoa tree, its pests and diseases, and its preparation for the market, has lagged behind research on crops such as rubber and tea because cocoa has not had plantation interests pressing for the results of research and also because of the nature of some of the problems involved. Research on cocoa has been and still is financed by the governments of cocoa-growing countries and to some extent by manufacturers, but cocoa growers generally do not contribute directly to the cost of research. This section is intended to describe briefly the various major centres of cocoa research and their development but will only indicate the lines of work followed as many of the findings have been described in more detail in other chapters.

A cocoa research unit was formed at the Imperial College of Tropical Agriculture, Trinidad, in 1930, at a time of crisis in the Trinidad cocoa-growing industry. Low prices, low yields and a new disease, witches' broom, had brought cocoa growers to bankruptcy. The cocoa research unit was financed by chocolate manufacturers and some cocoa growing countries and during the 1930s conducted the selection work which produced the 100 ICS clones, together with methods of vegetative propagation for their multiplication. After the war this unit was revived and strengthened, continuing its work of a long-term nature, on physiology, agronomy, plant breeding and the biochemistry of the cocoa bean. The unit continues to work on cocoa-breeding with the support of chocolate manufacturers in the United Kingdom and United States.

Cocoa research was started in Brazil as early as 1923 at a small experimental station at Uruçuca in the heart of the cocoa zone. This station was improved in 1931 under the Cocoa Institute of Bahia and became known for the discovery of Catongo cocoa, a white-seeded mutant of the Forastero types grown in Bahia. Catongo cocoa yields well and has been planted extensively.

In 1957 a new organisation, CEPLAC, was created, with sounder finances in the form of a levy on cocoa exports. This organisation covers research, extension and credit to cocoa growers, and in this way it is unique. The levy produces large sums which are adequate for the various tasks. The research wing, CEPEC, was established in 1963 at a new station near Itabuna. The effectiveness of the whole organisation is reflected in the rapidly increasing quantities of fertiliser and pesticides sold in the cocoa area.

Moving to West Africa research on cocoa was started as a result of the appearance of swollen shoot in Ghana in the late 1930s. A cocoa research station was established at Tafo in 1938 and this became the centre for British West Africa in 1944 when it became the West African Cocoa Research Institute (WACRI). Inevitably much of the work in the early years was concerned with virus disease but the introduction of planting material from Trinidad in 1944 had far-reaching consequences for plant breeding in West Africa. The work of WACRI included capsid studies which led to control of these pests with BHC, and agronomy in which the shade and fertiliser trial showed that unselected Amelonado could yield over 3,000 kg per hectare under suitable conditions.

This work was extended to Nigeria by the creation of a substation in 1953. Soon after independence was granted to Ghana and Nigeria WACRI became the Cocoa Research Institute for Ghana and the substation at Ibadan became the Cocoa Research Institute of Nigeria. These large institutes have research programmes covering all the main fields of crop research.

In the Ivory Coast and Cameroon cocoa research work is under the control of L'Institut Français du Café et du Cacao (IFCC). As its title implies this institute is responsible for research on coffee and cocoa, and deals also with tea and kola (Coste, 1966). It has its headquarters in Paris and is partly supported by the metropolitan government, partly by the governments of the countries in which it is working. IFCC started work in the Ivory Coast in 1959 and has laboratories at Bingerville and field stations at Abengourou and Divo. In 1964 work was extended to Cameroon where there are stations at Nkolbisson near Yaounde and at Nkoemvone where the field trials are conducted.

All these research units and institutes publish their work in annual reports as well as in papers in scientific journals. A major means of communication between scattered research workers and also extension workers has been the series of cocoa conferences that have been held since the war. Initially these conferences were held in London under the aegis of the Cocoa, Chocolate and Confectionery Alliance and were initiated in order to draw attention to the threat from virus disease. The latest of these conferences was held in 1961, since when a series of international cocoa research conferences have been held at roughly two-yearly intervals; the first was held at Abidjan in 1965, others have been held at Bahia, Accra and in Trinidad.

These conferences provide a meeting point for producers and manufacturers and the latter have made some significant contributions to cocoa research apart from supporting research in Trinidad. These contributions have included the work on capsids following the development of resistance to BHC and work on the epidemiology of black pod disease.

Manufacture

The essence of cocoa and chocolate manufacture lies in the development of flavour by roasting the beans followed by the extraction of cocoa butter from the nib to produce cocoa powder, and the addition of cocoa butter to nib and sugar to produce chocolate. The various steps in these processes are shown in a flow diagram (Fig. 15.1).

Initially the beans are cleaned to remove any foreign matter and to separate small or broken beans and clusters which will not roast uniformly with normal beans. The roasting process consists of heating the beans to 100 to 140°C for 45 to 90 minutes, the conditions varying according to the nature of the product. Roasting develops flavour and colour, and facilitates the removal of shell. Shell and nib (the roasted cotyledon) are separated in the next process of cracking or kibbling in a mill followed by winnowing. The shell is unsuitable for human consumption, but as complete separation is virtually impossible a tolerance of 1 to 2 per cent is allowed in food regulations. It is relatively easy to separate the larger pieces of shell and nib, but the finer particles cannot be completely separated, and a portion of shell–nib mixture is usually diverted for fat extraction. The larger fractions of shell are a

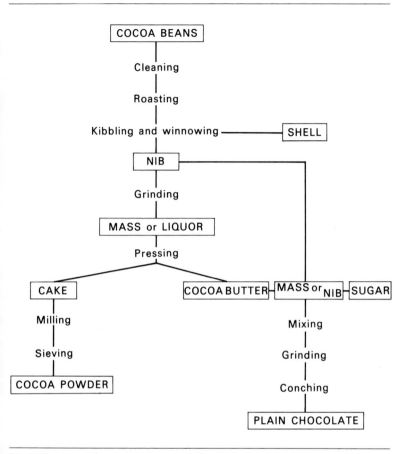

Fig. 15.1   Flow diagram of cocoa and chocolate production.

byproduct of relatively little value. The presence of theobromine in shell limits its use as a fodder.

The separated nib is ground to a liquor or mass which has a fat content of 55 to 58 per cent. For the production of cocoa powder some of the butter is removed by hydraulic presses. This will reduce the butter content to 22 to 23 per cent, which is normal for cocoa powder. The same presses can be used to produce cocoa powder of lower fat content down to 10 to 11 per cent but the process takes longer and, where butter production is the objective, an extrusion press will do the job more efficiently. This type of press can use whole beans or nib which have been conditioned by steaming before extrusion. To extract more butter it is necessary to resort to solvent extraction.

The cocoa powder is taken from the press as a cake, which has to be broken down in a mill. The resulting powder is sieved through fine silk, nylon or wire mesh. Most cocoa powders are made from mass which has been treated with alkali with the purpose of controlling the colour of the powder and improving its dispersability.

To produce plain chocolate nib or mass is mixed with sugar and sufficient cocoa butter to enable the chocolate to be moulded. The ratio of mass to sugar varies according to the national taste. The mixture is ground to such a degree that the chocolate is smooth to the palate. At one time this was done by a lengthy process in *melangeurs*— heavy granite rollers in a revolving granite bed—but nowadays grinding is done in a series of rolls. After grinding the chocolate is conched. The original conche was a tank shaped rather like a shell in which a roller is pushed to and fro on a granite bed. During the conching process which may last for several hours the chocolate is heated, this helps to drive off volatile acids, thereby reducing acidity when present in the raw bean, and the process finishes the development of flavour and makes the chocolate homogeneous.

Similar processes are involved in the manufacture of milk chocolate. The milk is added in various ways either in powder form to the mixture of mass, sugar and cocoa butter, or by condensing first with sugar, adding the mass and drying this mixture under vacuum. The product is called 'crumb' and this is ground and conched in a similar manner to plain chocolate.

After conching the chocolate has to be tempered before it is used for moulding or for enrobing confectionery centres. Tempering involves cooling and reaching the right physical state for rapid setting after moulding or enrobing.

Cocoa butter substitutes

The high prices for cocoa which prevailed after the war led to a search for suitable alternatives to cocoa butter. This is not a simple matter, as the particular mixture of fats which make up cocoa butter and which are vital for some of the attributes of chocolate is not easily contrived. The only natural substitute is Illipe butter which is extracted from seeds of *Shorea* species. These trees grow wild in the jungles of Sarawak, but the crop of nuts is irregular, so the supply of Illipe butter is similarly variable. Other substitutes have been manufactured from natural fats and have the same physical properties as cocoa butter.

The extent to which these substitutes are used depends on the relative prices of cocoa butter and the substitutes and on the proportion they can form of the finished product without altering the flavour. It is doubtful whether these substitutes replace more than 5 per cent of cocoa butter in any country.

Other uses for cocoa butter

The physical properties of cocoa butter, in particular its melting point at 34 to 35°C, i.e. just below body temperature, makes it suitable for certain cosmetics and pharmaceutical products, but the consumption of cocoa butter for these purposes is trivial in relation to the quantities used in chocolate manufacture.

The food value of cocoa and chocolate

Cocoa powder and chocolate contain a proportion of carbohydrates, fats and protein together with some vitamins of the vitamin B complex. Milk chocolate will contain, in addition, milk protein, calcium and other minerals and vitamins. Analyses of cocoa powder and chocolate are given in Tables 15.4 and 15.5. Both cocoa powder and chocolate have a high calorific value. Cocoa makes a very nutritious drink when mixed with milk and sugar and the presence of 2 per cent of theobromine gives a mildly stimulating action. Chocolate has excellent keeping qualities and is a safe concentrated food; the inclusion of milk protein in milk chocolate provides protein of high biological value. Chocolate is therefore used widely on expeditions and to a large extent as a snack between meals.

Table 15.4   *Analytical data for cocoa powder (mean of three varieties)*

Nutrients	g/100 g		Elements	mg/100 g
Protein	20·4		Sodium	650*
Fat	25·6		Potassium	534*
Available carbohydrate as				
monosaccharide	35·0		Calcium	51·2
			Magnesium	192
Calories per 100 g		452	Iron	14·3
			Copper	3·40
			Phosphorus	685 (15% of which is in the form of phytic acid phosphorus and not readily available)
			Sulphur	160
			Chlorine	199

*These values would be expected to vary with the manufacturing process used.
100 grams = 3·53 ounces.
SOURCE: McCance and Widdowson (1967).

Table 15.5 *Average composition of plain and milk chocolate*

No. of well-known varieties analysed	4 Plain	2 Milk
Total available carbohydrate†, calculated as monosaccharide		
(g/100 g)	52·5	54·5
Fat* (g/100 g)	35·2	37·6
Protein* (N × 6·25) (g/100 g)	5·6	8·7
Elements present (mg/100 g):		
Sodium	143·0	275·0
Potassium	257·0	349·0
Calcium	63·0	246·0
Magnesium	131·0	58·9
Iron	2·90	1·71
Copper	0·81	0·49
Phosphorus	138·0	218·0
Chlorine	4·8	170·0
Calories per 100 g	544·0	588·0

*The total fat and protein in milk chocolate will include a proportion of milk fat (the other fat normally being cocoa butter), and of milk protein.
†Total available carbohydrate will consist of sugar, cocoa starch and also lactose in the case of milk chocolate.
SOURCE: McCance and Widdowson (1967).

References

Alba, J. de and Basadre, J. (1952) 'Trials of fattening pigs with rations based on cacao pod meal, maize and ripe bananas', *Turrialba*, **2** (3), 106–9.

Alba, J. de, Garcia, H., Cano, F. P. and Ulloa, G. (1954) 'Nutritive value of cacao pods in milk production in comparison with ground maize and cassava meal', *Turrialba*, **4** (1), 29–34.

Coste, R. (1966) 'L'Institut Français du Café et du Cacao', *Cocoa Growers' Bull.*, **7**, 9–11.

Greenwood-Barton, L. H. (1965) 'Cocoa beans and cocoa pods; have they any unconventional uses?', *First Session FAO Tech. Wkg Party Cocoa Prodn, Rome 1964*, Paper Ca. 64/7.

McCance, R. A. and Widdowson, E. M. (1967) *The Composition of Foods*, Medical Res. Council, Spec. Rep. Series No. 297.

Owusu-Domfeh, K. (1972) 'The future of cocoa and its by-products in the feeding of livestock', *Ghana J. Agric. Sci.*, **5**, 57–64.

Owolabi, C. A. (1972) 'Cocoa quality survey', *Ann. Rep. Cocoa Res. Inst. Nigeria 1969–70*.

Chapter 16

Labour Requirements and Costs

Detailed statistics on costs are available for a very limited number of cocoa estates and from a few surveys of smallholdings in West Africa. Figures have been published for an estate in West Cameroon and one in Malaysia, in both instances the cocoa was established under thinned jungle (Wood, 1961, 1966, 1974). In both estates the cost of production was generally between £150 and £200 per ton and, as one would expect, was highly dependent on the level of yield. Yields of about 800 kg per hectare were the minimum necessary to make a profit under the conditions of cost and prices prevailing during the 1960s.

The establishment of cocoa under coconuts is attractive because of the low capital cost involved and the additional income it generates. Calculations based on estate experience have been published (Leach, 1968; Recter *et al.*, 1972).

The detailed figures given in these papers is not repeated here because the information is generally of interest at the time of publication but rapidly becomes out of date as costs rise and exchange rates change. The papers quoted do, however, contain data on labour requirements which are of more lasting value and these form the basis of this chapter.

Establishment

The figures in Table 16.1 give the man-days per acre in a succession of annual plantings at a plantation in West Cameroon. The cocoa was planted under thinned jungle, which was first underbrushed, clearing all the undergrowth and creepers, after which the blocks were lined and staked. Then the jungle was thinned so as to achieve the degree of shade required. The fallen trees were cut up so that the planting lines could be cleared. The spacing was 3 by 3 m (10 by 10 ft) and the planting holes were made just large enough for the seedling and its ball of earth a short time ahead of planting. The planting season started in April and planting was completed in June. The table shows that labour requirements for this method are 90 to 100 man-days per acre for the initial

Table 16.1   *Establishment at Ikiliwindi, West Cameroon (average figures, man-days per acre)*

Task	105 acres planted in 1958			202 acres planted in 1959			210 acres planted in 1960		
	1st	2nd	3rd	1st	2nd	3rd	1st	2nd	3rd
Underbrushing	17·0	—	—	14·0	—	—	12·0	—	—
Staking	17·0	—	—	12·5	—	—	10·0	—	—
Felling	21·0	—	—	23·5	—	—	26·0	—	—
Holing	4·5	—	—	4·0	—	—	4·0	—	—
Planting	6·0	—	—	5·0	—	—	5·0	—	—
Nursery	15·0	—	—	15·0	—	—	15·0	—	—
Shade planting	0·5	—	—	1·0	—	—	0·5	—	—
Tree killing	—	—	0·5	—	—	0·5	—	—	0·5
Spraying	1·5	2·0	3·0	1·5	1·0	3·0	1·0	—	2·0
Weeding	21·0	27·0	37·5	22·0	25·5	26·0	22·5	15·5	24·0
Supplying	0·5	2·0	—	—	—	2·0	—	2·0	1·0
	104·0	31·0	41·0	98·5	26·5	31·5	96·0	17·5	27·5

SOURCE: Estate records.

preparation of the land and the planting year, up to eighteen months might be spent in all, and 20 to 30 man-days per acre in the second and third years. After the third year the cocoa started to bear and the plantings were regarded as mature for accounting purposes; the time at which this change takes place is an arbitrary one and will differ with the planting material and the growing conditions as well as the tax laws of the country concerned.

Establishment after clear-felling is inevitably more costly. Data from Fiji (Harwood, 1959) showed a labour requirement of 119 man-days per acre for first year excluding nursery costs. A study on costs in New Guinea (Shaw, 1958) indicated that the first year—preparation of the land, planting and weeding—would require 100 man-days per acre for rough clearing and up to 240 man-days per acre for clean clearing.

Figures for land clearing in Ghana have been given by Allison and Cunningham (1959), who described a special case when land was prepared for a field experiment which involved cutting and removing of all fallen logs; in this case the preparation of the land and the planting of shade and cocoa required 293 man-days per acre of which 226 man-days were used in felling, cutting and removal of logs. Various alternative methods have been described by Allison and Smith (1964).

Total cost of a cocoa plantation

The total costs comprise field costs, cost of buildings and equipment and overheads or general charges. This last item consists of the cost of

Table 16.2 Labour requirements for maintaining cocoa, Ikiliwindi, West Cameroon

Task/Cost	126 acres planted in 1958					323 acres planted in 1961				
	1966–67	1967–68	1968–69	1969–70	1970–71	1966–67	1967–68	1968–69	1969–70	1970–71
	Man-days per acre									
Tree killing	—	—	0·2	—	0·2	—	—	0·1	—	0·2
Spraying	9·8	8·0	9·8	14·7	11·2	6·4	5·0	8·9	12·4	10·3
Weeding	8·2	10·0	6·0	5·4	5·0	9·4	12·0	8·6	4·8	7·5
Supplying	1·6	1·0	1·2	0·8	0·1	2·2	1·5	1·7	0·2	0·3
Pruning	1·9	1·5	0·8	2·5	0·9	1·0	—	—	3·5	—
Harvesting	11·9	13·0	10·7	8·1	13·4	8·7	8·0	11·0	9·7	13·7
TOTAL MAN-DAYS	33·4	33·5	28·7	31·5	30·8	27·7	20·5	30·3	30·6	32·0
	£ per acre									
Cost: Labour	9·70	10·75	11·55	11·50	11·90	8·35	10·00	12·10	11·25	12·10
Materials	6·10	6·20	10·85	5·90	7·96	5·75	4·65	7·25	4·90	6·62
TOTAL COST	£15·80	£16·95	£22·40	£17·40	£19·86	£14·10	£14·65	£19·35	£16·15	£18·72

SOURCE: Wood (1974).

management, transport, office expenses, rent and rates, health and welfare and upkeep of roads and buildings. The three items—planting, buildings, general charges—are roughly equal and in the case of the plantation in West Cameroon which was established between 1956 and 1963 the total cost was £170 per acre of cocoa. This figure would be considerably exceeded now.

Production costs

Field costs

Data from the plantation in West Cameroon are given in Table 16.2. This indicates a labour requirement of roughly 30 man-days per acre. In this area black pod disease is severe and 10 or more man-days have to be spent on spraying, a considerable proportion of the total. Data from a survey of estates in Trinidad are given in Table 16.3. These show a lower labour requirement which is inevitable when yields are rather low and labour costs high. Figures from Malaysia (Table 16.4) give a

Table 16.3   *Labour requirements on cocoa estates, Trinidad*

Task	Man-days per acre
Weed control	5·2
Pruning	3·6
Pests and diseases	2·5
Other maintenance tasks	4·2
Harvesting	3·5
Processing	4·0
Total	23·0

SOURCE: Lass, 1969.

Table 16.4   *Field costs per acre on an estate in Malaysia*

Task	1969–70 £	1970–71 £	1971–72 £
Weeding	2·75	2·38	2·72
Shade	1·14	1·40	1·44
Pests and diseases	4·40	3·40	4·45
Roads, drains, bunds	1·40	1·61	1·54
Fencing, pruning	0·90	1·00	1·08
Manuring	2·75	4·10	4·25
Experiments	0·32	0·95	1·00
Total field costs	£13·66	£14·84	£16·48

SOURCE: Wood, (1974).

similar total cost per acre but a very different division between the various items.

Fermentation and drying

Some figures of the cost of fermentation and drying in Cameroon and Malaysia are given in Table 16.5. While the cost of estate processing

Table 16.5 *Costs of fermentation and drying in Cameroon and Malaysia (£ per ton)*

Year	Cameroon	Malaysia
1968–69	23·32	—
1969–70	23·8	12·80
1970–71	24·9	13·90
1971–72	—	14·20

SOURCE: Wood (1974).

was appreciably higher in Cameroon it should be noted that the pods were opened centrally in Cameroon and the cost of pod opening is included in the table; on the estate in Malaysia the pods were opened in the field so the cost of pod opening is included in harvesting. In Cameroon the beans were fermented in trays and dried on platforms heated by oil-fired heat exchangers; in Malaysia box fermentation was practised and the beans dried on large Samoan dryers with sliding roofs. Looking at the individual steps in preparing cocoa for the market the first task is to harvest the pods and open them. These jobs can be organised in a variety of ways but the data from various sources suggest that the daily task of harvesting or opening is 1,500 to 2,000 pods. Using a pod value of 12, i.e. 12 pods to 1 lb of dry beans, there will be 27,000 pods to 1 ton and the labour requirement for harvesting and pod opening will be 27 to 36 man-days per ton. The labour involved in fermentation by any method is relatively small, about 1 man-day per ton. It is therefore surprising how much thought has been given to, and

Table 16.6 *Factory costs, Cameroon (£ per ton)*

	1966–67	1968–69	1970–71
Labour: pod opening	3·21	5·39	6·2
Labour: other	4·06	4·90	4·5
Fuel	6·90	9·05	10·6
Bags and liners	3·07	3·98	3·6
Total cost per ton	17·24	23·32	24·9

SOURCE: Wood (1974).

money spent in arranging fermentation boxes on slopes or in tiers. While this will save labour the amount saved is only 1 man-hour per ton of dry beans according to Rohan (1963). Some detailed costs of drying are given in Table 16.6.

Costs of production on smallholdings

The classic study of the costs and labour requirements on small farms in Ghana is Beckett's (1944) monograph on Akokoaso. This is based on a survey conducted in the 1930s and the data collected gave the following labour requirements:

	Man-days/acre/annum
Weeding	8·2
Harvesting	15·6
Carrying	1·4
Other	0·6
Total	25·8

These figures relate to farms yielding 340 lb per acre and the labour involved in producing 1 ton of dry beans was therefore 170 man-days.

A more general survey conducted in Nigeria in 1951–52 (Galletti *et al.,* 1956) showed that at a yield of 300 lb per acre labour requirements amounted to 31·5 man-days per acre, equivalent to 235 man-days per ton.

A recent survey (Okali, 1974) of a small number of farms in Ghana has given a figure of 27 man-days per acre, which is very close to the data from Akokoaso, although the divisions between the various operations is different. The details are:

	Man-days/acre/annum
Weeding	12
Other maintenance	2
Harvesting, breaking, drying	13
Total	27

The yields from these farms was 350 lb per acre, almost exactly the same as the earlier figure.

The 'capital' expenditure on smallholdings is almost entirely confined to the labour involved in establishing the farm and building a simple house. Equipment consists of tools, planting material, and possibly a spraying machine. In maintaining bearing farms the only

expenditure in addition to the labour already detailed will be on
pesticides to control capsids or black pod disease.

References

Allison, H. W. S. and Cunningham, R. K. (1959) 'Preparing land for cacao in Ghana',
World Crops, **11**, 311–13.

Allison, H. W. S. and Smith, R. W. (1964) 'Economics of cacao establishment on clear-
felled land', *World Crops*, **16**, 31–6.

Beckett, W. H. (1944) *Akokoaso: a survey of a Gold Coast village*, Monographs of
Social Anthropology No. 10, London.

Galletti, R., Baldwin, K. D. S. and Dina, I. O. (1956) *Nigerian Cocoa Farmers: an
economic study of Yoruba cocoa farming families*, Oxford University Press.

Harwood, L. W. (1959) 'Cocoa planting', *Agric. J. Fiji*, **29**, 65–74.

Lass, R. A. (1969) 'Some aspects of the economics of cocoa production in Trinidad and
Tobago', unpublished thesis, Univ. West Indies.

Leach, J. R. (1968) 'The economics of cocoa–coconut interplanting', *Cocoa and Coconuts
in Malaya. Symp. Incorp. Soc. Planters, Kuala Lumpur 1967*, pp. 103–11.

Okali, C. (1974) 'Labour inputs on cocoa farms: case studies', *Economics of cocoa
production and marketing. Cocoa Economics Research Conf.*, Univ. of Ghana, 1973,
pp. 3–19.

Recter, D. H., Leach, J. R. and Lim Kiam Peng (1972) 'Cost of production and expected
return from cocoa planted under coconuts', *Cocoa and Coconuts in Malaysia. Proc.
Conf. Incorp. Soc. Planters, Kuala Lumpur 1971*, pp. 251–68.

Rohan, T. A. (1963) *Processing of Raw Cocoa for the Market*, FAO Agric. Studies
No. 60.

Shaw, D. D. (1958) *Cocoa in Papua and New Guinea I. An economic study of the cocoa
growing industry in Papua and New Guinea*, Canberra.

Wood, G. A. R. (1962) 'Costs of cocoa production', *Rep. Cocoa Conf. London 1961*,
pp. 36–50.

Wood, G. A. R. (1966) 'Further costs of cocoa production', 2nd session, FAO Tech. Wkg
Pty Cocoa Production and Protection, Rome, Paper Ca/66/15.

Wood, G. A. R. (1974) 'Some aspects of cocoa production costs on plantations', *Economics of cocoa production and marketing. Cocoa Economics Research Conf.*, Univ. of
Ghana, 1973, pp. 60–8.

Appendix 1

International Cocoa Standards

The following is the text of the Model Ordinance and the Code of Practice:

Model Ordinance

1. Definitions

Cocoa bean: The seed of the cocoa tree (*Theobroma cacao* Linnaeus); commercially and for the purpose of this Model Ordinance the term refers to the whole seed, which has been fermented and dried.

Broken bean: A cocoa bean of which a fragment is missing, the missing part being equivalent to less than half the bean.

Fragment: A piece of cocoa bean equal to or less than half the original bean.

Piece of shell: Part of the shell without any of the kernel.

Adulteration: Alteration of the composition of graded cocoa by any means whatsoever so that the resulting mixture or combination is not of the grade prescribed, or affects injuriously the quality or flavour, or alters the bulk or weight.

Flat bean: A cocoa bean of which the cotyledons are too thin to be cut to give a surface of cotyledon.

Foreign matter: Any substance other than cocoa beans, broken beans, fragments, and pieces of shell.

Germinated bean: A cocoa bean, the shell of which has been pierced, slit or broken by the growth of the seed germ.

Insect-damaged bean: A cocoa bean the internal parts of which are found to contain insects at any stage of development, or to show signs of damage caused thereby, which are visible to the naked eye.

Mouldy bean: A cocoa bean on the internal parts of which mould is visible to the naked eye.

Slaty bean: A cocoa bean which shows a slaty colour on half or more of the surface exposed by a cut made lengthwise through the centre.

Smoky bean: A cocoa bean which has a smoky smell or taste or which shows signs of contamination by smoke.

Thoroughly dry cocoa: Cocoa which has been evenly dried throughout. The moisture content must not exceed 7·5 per cent.[1]

2. Cocoa of Merchantable Quality

(a) Cocoa of merchantable quality must be fermented, thoroughly dry, free from smoky beans, free from abnormal or foreign odours and free from any evidence of adulteration.
(b) It must be reasonably free from living insects.
(c) It must be reasonably uniform in size,[2] reasonably free from broken beans, fragments, and pieces of shell, and be virtually free from foreign matter.

3. Grade Standards

Cocoa shall be graded on the basis of the count of defective beans in the cut test. Defective beans shall not exceed the following limits:

Grade I —(a) mouldy beans, maximum 3 per cent by count;
       (b) slaty beans, maximum 3 per cent by count;
       (c) insect-damaged, germinated, or flat beans, total maximum 3 per cent by count.

Grade II—(a) mouldy beans, maximum 4 per cent by count;
       (b) slaty beans, maximum 8 per cent by count;
       (c) insect-damaged, germinated, or flat beans, total maximum 6 per cent by count.

Note: When a bean is defective in more than one respect, it shall be recorded in one category only, i.e. the most objectionable. The decreasing order of gravity is as follows:
—mouldy beans;
—slaty beans;
—insect-damaged beans, germinated beans, flat beans.

---

[1] This maximum moisture content applies to cocoa in trade outside the producing country, as determined at first port of destination or subsequent points of delivery.
The Working Party reviewed the ISO method for determination of moisture content and agreed that it could be used, when recommended by ISO, as a practical reference method.

[2] *Uniform in size:* As a guide not more than 12 per cent of the beans should be outside the range of plus or minus one-third of the average weight. It is recognised, however, that some hybrid cocoa may not be able to meet this standard although fully acceptable to the trade.

4. Sub-standard cocoa

All dry cocoa which fails to reach the standard of Grade II will be regarded as sub-standard cocoa and so marked (SS), and shall only be marketed under special contract.

5. Marking and Sealing

(a) All cocoa graded shall be bagged and officially sealed. The bag or seal shall show at least the following information:
Producing country, grade or 'SS' if sub-standard, and whether light or mid-crop,[1] and other necessary identification marks in accordance with established national practice.[2]
(b) The period of validity of the grade shall be determined by governments in the light of climatic and storage conditions.

6. Recheck at Port of Shipment

Notwithstanding paragraph 5 (b) above, all cocoa so graded shall be re-checked at port within seven days of shipment.

7. Implementation of Model Ordinance

Methods of sampling, analysis, bagging, marking and storage applicable to all cocoa traded under the above International Standards are set out in the following Code of Practice.

Code of Practice

A. Inspection

1. Cocoa shall be examined in lots, not exceeding 25 tons in weight.
2. Every parcel of cocoa shall be grade-marked by an inspector, after determining the grade of the cocoa on the basis of the cut test (see paragraph C below).
Grade marks shall be in the form set out in, and shall be affixed according to, Section ... of ...[3] and shall be placed on bags by means of a stencil or stamp (see also paragraph E below).

---

[1] Absence of a crop indication means main crop.
[2] Some members of the Working Party felt that it was desirable that the cocoa year also be shown. The Working Party therefore decided that this question should be re-examined at its Fourth Session and that in the meantime indication of the crop year was optional.
[3] i.e. the appropriate reference in national regulations.

B. Sampling

1. Samples for inspection and analysis should be obtained:
   (a) from cocoa in bulk, by taking samples at random from the beans as they enter a hopper or from the top, middle and bottom of beans spread on tarpaulins or other clean, dust-free surface, after they have been thoroughly mixed;
   (b) from cocoa in bags by taking samples at random from the top, middle and bottom of sound bags using a suitable stab-sampler to enter closed bags through the meshes of the bags, and to enter unclosed bags from the top.
2. The quantity of samples to be taken should be at the rate of not less than 300 beans for every ton of cocoa or part thereof, provided that in respect of a consignment of one bag or part thereof, a sample of not less than 100 beans should be taken.
3. For bagged cocoa, samples shall be taken from not less than 30 per cent of the bags, i.e. from one bag in every three.
4. For cocoa in bulk, not less than five samplings shall be taken from every ton of cocoa or part thereof.
5. In importing countries samples for inspection should be taken from not less than 30 per cent of each lot of 200 tons or less, i.e. from one bag in three. Samples should be taken at random from the top, the middle and the bottom of the bag.

C. The Cut Test

1. The sample of cocoa beans shall be thoroughly mixed and then 'quartered' down to leave a heap of slightly more than 300 beans. The first 300 beans shall then be counted off, irrespective of size, shape and condition.
2. The 300 beans shall be cut lengthwise through the middle and examined.
3. Separate counts shall be made of the number of beans which are defective in that they are mouldy, slaty, insect damaged, germinated, or flat. Where a bean is defective in more than one respect, only one defect shall be counted, and the defect to be counted shall be the defect which occurs first in the foregoing list of defects.
4. The examination for this test shall be carried out in good daylight or equivalent artificial light, and the results for each kind of defect shall be expressed as a percentage of the 300 beans examined.

D. Bagging

1. Bags should be clean, sound, sufficiently strong and properly sewn. Cocoa should be shipped only in new bags.

E. Sealing and Marking

1. After grading, each bag should be sealed with the individual examiner's seal. The grade should be clearly marked on each bag. Bags should also be clearly marked to show the grading station and period of grading (week or month).

   For these purposes the following measures shall be carried out:

   (a) suitable precautions will be taken in the distribution and use of examiners' seals to ensure that they cannot be used by any unauthorised person;

   (b) parcels shall be numbered consecutively by the official examiner with lot numbers from the beginning of each month. The parcel number or lot number will be stencilled on each bag in every parcel examined, in the corner nearest the seal;

   (c) grade marks will be stencilled near the mouth of the bag.

F. Storage

1. Cocoa shall be stored in premises constructed and operated with the object of keeping the moisture content of the beans sufficiently low, consistent with local conditions.

   Storage shall be on gratings or deckings which allow at least 7 cm of air space above the floor.

2. Measures shall be taken to prevent infestation by insects, rodents and other pests.

3. Bagged cocoa shall be so stacked that:

   (a) each grade and shipper's mark is kept separate by clear passages of not less than 60 cm in width, similar to the passage which must be left between the bags and each wall of the building;

   (b) disinfestation by fumigation (e.g. with methyl bromide) and/or the careful use of acceptable insecticide sprays (e.g. those based on pyrethrin) may be carried out where required; and

   (c) contamination with odours or flavours or dust from other commodities, both foodstuffs and materials such as kerosene, cement or tar, is prevented.

4. Periodically during storage and immediately before shipment, the moisture content of each lot should be checked.

G. Infestation

1. Cocoa beans may be infested with insects which have not penetrated the beans and whose presence is not revealed by the cut test which is employed for grading purposes. Such insects may subsequently enter beans or they may be involved in cross infestation of other shipments.

2. Therefore, when the cocoa is rechecked at port before shipment, as

provided under paragraph 6 of the Model Ordinance, it should also be inspected for infestation by major insect pests. If it is found to be infested it should, before shipment, be fumigated, or otherwise treated to kill the pests. Care should be taken to avoid cocoa beans becoming infested in ships and stores from other commodities or with insects remaining from previous shipments.

3. If the use of insecticides or fumigants is necessary to control infestation, the greatest care must be exercised in their choice and in the technique of their application to avoid incurring any risk of tainting or the addition of toxic residues to the cocoa. Any such residues should not exceed the tolerances prescribed by FAO/WHO Codex Committee on Pesticide Residues and the FAO/WHO Expert Committee on Pesticide Residues and by the government of the importing country.

4. Rodents should as far as possible be excluded from cocoa stores by suitable rodent-proof construction, and where direct measures are necessary to control rodents the greatest care must be taken to prevent any possibility of contaminating the cocoa with substances which may be poisonous.

Appendix 2

Conversion Factors

The figures in the left hand column are used in converting the factor on the right to the factor on the left, i.e. in the direction of the arrow; and vice versa for the figure in the right hand column.

	←	→	
Centimetres	2·54	0·394	Inches
Metres	0·305	3·28	Feet
Hectares	0·405	2·471	Acres
Kilogrammes	0·454	2·205	Pounds
Grammes	28·35	0·035	Ounces
Metric tons	1·016	0·984	Long tons
Litres	4·546	0·22	Gallons
Litres per hectare	11·21	0·089	Gallons per acre
cc per 100 litres	6·24	0·16	Fluid oz per 100 gal
Kilogrammes per hectare	1·121	0·892	Pounds per acre
Cubic metres	0·28	35·3	Cubic feet

Appendix 3

Visual Symptoms of Mineral Malnutrition

A. Symptoms more or less general on the whole plant.
Element deficient—nitrogen, sulphur, phosphorus.
Element toxic—boron.

B. Symptoms confined to, or at least more pronounced in, the older leaves.
Element deficient—calcium, magnesium, potassium.
Element toxic—aluminium, chlorine, iron.

C. Symptoms confined to, or more pronounced in, the younger leaves.
Element deficient—iron, manganese, copper, zinc, boron, molybdenum.
Element toxic—zinc, manganese, copper.

A. 1. Leaves pale or yellowish in colour, reduced in size, older leaves finally showing tip scorch: younger leaves small, yellow or almost white in colour with little or no green associated with the veins: internodes compressed and petioles showing acute angle with stem.—Nitrogen deficiency (Plate II).
2. Leaves of whole plant pale yellowish or yellowish green in colour, but no marked reduction in size. Yellow blotches on older leaves. New flush leaves normal in size, at first bright yellow in colour with no green associated with the veins, later becoming pale yellowish green as in older leaves: plant frequently preferentially attacked by insect pests.—Sulphur deficiency.
3. Plant somewhat stunted in growth: mature leaves paler towards tip and margin, followed by tip and marginal scorch. Young leaves markedly reduced in size, often showing interveinal pallor; stipules frequently persisting after leaf abscission; young leaves showing acute angle with stem, internodes compressed.—Phosphorus deficiency.
4. Older leaves showing pronounced marginal scorch and necrotic areas in vicinity of wounds, younger leaves cupped downwards,

showing green in vicinity of veins, with broad chlorotic inter-veinal areas later greening slightly and developing necrotic tip and margin.—Boron toxicity.

B. 1. Necrotic areas commencing in interveinal region near leaf margin quickly fusing into continuous marginal necrosis of older leaves. No necrotic lesions in advance of main marginal necrotic zone, unaffected area showing oak leaf pattern.— Calcium deficiency (Plate V).

  2. Necrotic areas commencing in interveinal region near leaf margin, quickly fusing into continuous marginal necrosis, of older leaves. Prominent bright yellow zone in advance of necrotic area and islands of necrotic tissue often appearing in advance of main wave of necrotic tissue. Unaffected areas of the leaf paler green than usual and forming oak leaf pattern.— Magnesium deficiency.

  3. Pale yellow areas formed in interveinal region near leaf margin, quickly becoming necrotic but only fusing with each other after some time; progress of marginal necrosis much more rapid between veins; yellow zone on inner surface of invading necrotic zone. Potassium, calcium and magnesium deficiencies are not easy to differentiate in the field. —Potassium deficiency (Plate IV).

  4. Paler or yellowish areas in the interveinal region of the distal end (tip) of leaf with tip scorch progressing very slowly. Rarely, some blackening of the interveinal region towards the base of the leaf. All symptoms confined to older leaves only.—Aluminium toxicity.

  5. Pale yellow areas developing in the marginal interveinal regions quickly fusing to form a continuous scorch, advancing more rapidly in interveinal areas. Tissues in advance of scorched area showing various shades of dark green and grey. Scorch pro-ceeding slowly and necrotic areas in vicinity of wounds. Can be confused with the calcium deficiencies above.—Chlorine toxicity.

  7. Pale yellow zone on each side of the midrib of older leaves, rapidly spreading necrotic areas formed in vicinity of wounds; no marginal or tip necrosis.—Iron toxicity.

C. 1. Younger leaves showing darker green veins against paler green background, or showing green tinted veins against pale yellowish white or almost completely white background: developing tip scorch. Symptoms less marked in leaves of previous flush—older leaves frequently showing narrow marginal and tip scorch.— Iron deficiency (Plate III).

  2. Younger leaves pale yellowish or yellowish green, later develop-ing blurred chlorotic pattern in which the tissues in the vicinity of the midrib, main laterals and tertiary veins are prominently

green against pale background; followed by scorch primarily of the tip and distal margin.—Manganese deficiency.

3. Leaves on young flush small but normal in shape—young shoots frequently showing signs of wilting. Sudden collapse of tissues at tip of leaf—collapsed tissues remaining green for some time, later forming brown edge with apex directed towards midrib. No marked chlorotic pattern.—Copper deficiency.

4. Very young leaves showing prominent dark red veinlets with considerable distortion, leaf very narrow in proportion to length, margin often wavy and leaf sometimes sickle-shaped with small chlorotic patches in distinct row on each side of midrib only or on each side of midrib and main lateral veins.—Zinc deficiency.

5. Young leaves reduced in size, pale, hardening with marked reflexed curvature and/or spiral twisting, thick to the touch and brittle; old leaves of healthy appearance.—Boron deficiency.

6. Young leaves thin and translucent, developing mild chlorotic mottling more marked in interveinal region, later developing marginal scorch.—Molybdenum deficiency.

7. Young leaves showing olive green appearance or pale green areas scattered over surface of leaf.—Zinc toxicity.

8. Youngest mature leaves showing irregular pale green or yellowish areas on darker green background with or without some veinal necrosis; no tip or marginal scorch and no symptoms on older leaves.—Manganese toxicity.

9. Young leaves showing dark olive green colour with upraised veinlets and puckering of lamina along midrib. Younger mature leaves showing pale green areas distributed at random over leaf surface.—Copper toxicity.

Index

References to figures are shown in *italic*,
plates are shown in **bold** type.

284

Endoclyta (= *Phassus*) *hosei* (borer), 177–8

Endosulfan, 173, 191

Endrin, 164, 169, 180–1

Enxerto ants, 183–4

Erythrina sp., 77–9, 112; **28**, **29**

Estufa dryer, 221

Eugenia (windbreak), 114

Eulophonotus (wood boring moth), 164

F

Fan-branches, definition, 9

Fan gall, 143

Fat content, 231–4

Fenitrothion, 173, 199

Fermentation, 14, 204–17

basket, 213

box, 206–9, 212–13; **65**, **66**

heap, 206–8, 213; **63**, **64**

interrupted, 216

pH, *12.4*, 207

temperature curve, *12.2*, 206, *12.3*, 207

tray, 213–14; **67**

Fernando Po, 2, 16, 19, 27, 52, 97, 136–7, 184, 220

Fertilisers, 67, 107–10, 116–23, 135

Fiji, 27, 86, 89, 184, 268

Fine grade cocoas, 4, 16, 48, 230, 241, 250

Flat beans, 238, 274

Flavour, *see* chocolate flavour

Flavour cocoas, *see* fine grade cocoas

Flower, description, 11–12, *2.2*

Flowery gall, 143

Flushing, 10, 29

Forastero cocoa, 1, 4, 14–17, 48–9, 55–8, 232, 242

diseases, 131, 145, 147

fermentation, 207, 210–11

pests, 180

Forcipomyia sp., 12–13

Free fatty acids, 210, 236

Fungicide, seed treatment, 61

Fusarium rigidiuscula, 142

Fusarium roseum, 145

G

Galba (*Calophyllum antillanum*), 86, 114

Gamma—BHC, 169. *See also* BHC

Germinated beans, 210, 238, 274

Germination, 7–8

Ghana

climate, 19, *3.1*, 28

cocoa, 16, 48, 59, 65, 71, 80, 93, 98, 101–3, 122, 201, 242, 261

breeding, 52–4

costs, 268, 272

diseases, 126–30, 133–4, 142, 144, 151, 153–7

establishment, 82, 84, 87

fermentation, 213–14

fertilisers, 109, 121

pests, 164, 169–72, 174, 180, 191

production, 252–5

quality, 232–4, 245

Glenea aluensis, 187

Glenea lefebueri, 187

Glenea novemguttata, 187

Gliricidia sp., 69, 80

Gliricidia sepium (=*maculata*), 78, 112; **13**, **27**

Grafting, 71–2

Green-point gall, 143; **52**

Grenada, 68, 86, 90, 114, 139, 185, 200

Grinding, 264

Growth habit, 8, *2.1*

Guadeloupe, 185

Guatemala, 1, 143, 145

Guyana, 139, 174

H

Hand pollination, 60

Harmattan, 31, 148, 239

Harvesting, 202–4; **60**

Hawaii, 145

Heap fermentation, 206–8, 213; **63**, **64**

Heat exchangers, 220, 222

'Hedging', 247

Helopeltis sp., 166, 170; **57**

Helopeltis clavifer, 166, *11.4*

Heptachlor, 169

Herbicides, 94–7

Hibiscus (windbreak), 86, 114

Holing, 89–90

Horiola picta, 174

Horsehair blight, 153

Hybrid vigour, 57

I

Illipe butter, 264

Immortelle (*Erythrina* sp.), 77, 112–14

Imperial College Selections (I.C.S.), 51–2, 59, 69

Incompatibility, 12–13, 59

India, 99

Indonesia, 17, 58, 79, 232, 257. *See also* Java

Infestation, 237–8, 278–9

Infested beans, 237–8, 274–5

Inga sp., 112

Inga laurina, 78

Insecticides, list of, 199. *See also* individual insecticides